T0137106

Biomechanics for Instructors

Nikolai Aleksandrovich Bernstein

Biomechanics for Instructors

Translated by Rose Whyman

 Springer

Nikolai Aleksandrovich Bernstein
Moscow, Russia

Translated by Rose Whyman
Department of Drama and Theatre Arts
University of Birmingham
Birmingham, UK

Translated by Rose Whyman with an introductory essay by Vera L. Talis. This book was first published in Russian by Novaya Moskva in 1926 as a set of lectures for courses for instructors in the industrial-economic subdivision of Moscow Professional Training.

ISBN 978-3-030-36165-5   ISBN 978-3-030-36163-1   (eBook)
https://doi.org/10.1007/978-3-030-36163-1

This Springer imprint is published by the registered company Springer Nature Switzerland AG
The registered company address is: Gewerbestrasse 11, 6330 Cham, Switzerland

For Brian Door, as ever, with all my love and thanks.

Preface

I hope this translation of Bernstein's *Biomechanics for Instructors* will be of interest to people involved in a wide range of research and practice. It must be said that this set of lectures was developed and published in the 1920s, and N.A. Bernstein conveyed the state of biomechanics as it was known at that time. However, as is the case with any science, knowledge has advanced. Readers are encouraged to expand their study in this field by reviewing the latest findings in the biomedical literature and current textbooks in order to place Bernstein's explanations in the context of current knowledge. This is especially important with respect to discussions of physiology—whether humans or animals—as knowledge in this field has grown substantially over the last century.

This proviso does not diminish the significance of these lectures and Bernstein's work in general, which, as Vera Talis explains in the introductory essay, is far from being fully acknowledged or assessed. She details much of the valuable work that has been carried out by others but does not mention the tremendous work she herself has done to preserve Bernstein's legacy. Hopefully, her book *Doktor kotoryi lyubul parovoziki* (*The Physician Who Loved Steam Engines*) will soon be in print and much more information about Bernstein's life and work will be available to a general readership. Also, to date there has been a lack of recognition of how Bernstein's research impacted in fields beyond that of biomechanics and neurophysiology. He was influential in Soviet sport and athletics training, and much research into this influence and its impact on the development of skill and virtuosity in this area remains to be done.

In drama, dance and theatre work, we are beginning to understand his influence in his time and the significance of Bernstein's discoveries about movement in the period of emergence of the Free Dance Movement at the beginning of the twentieth century on which cultural historian Irina Sirotkina has carried out extensive and groundbreaking research, resulting in many articles and the 2012 book *Svobodnoye Dvizheniye i Plasticheskii Tanets v Rossii*, Moscow: Novoye Literaturnoye Obozreniye.

Irina Sirotkina's and science historian Roger Smith's *The Sixth Sense of the Avant-Garde Dance, Kinaesthesia and the Arts in Revolutionary Russia* (2017) includes discussion of Bernstein's work to show how significant movement and the sense of movement were to pioneers of modernism at the turn of the twentieth century, in revolutionary Russia and beyond.

Irina's *Mir Kak Zhivoe Dvizhenie- intellektual'naya biografia Nikolaya Bernshteina* (Moscow: Kogito-Tsentr 2018) is a further important contribution.

My own research encompasses the work of the world-famous theatre directors K.S. Stanislavsky and V.E. Meyerhold, and I am currently investigating how Meyerhold and others including the film director Sergei Eisenstein developed actor-training systems which they called Biomechanics and connections between these, and biomechanics. What is clear is that Bernstein's insight into the science of movement was profound, and a clearer understanding of this, prevented as Bernstein's persecution began, as described in the introductory essay would have been of use in actor training then and could be of great use in developing the practice of performer training today.

Finally, I would like to mention that I first heard of Bernstein during my training with the Professional Association of Alexander Teachers. In our studies of Movement Science, my teacher Brian Door introduced us to *The Co-ordination and Regulation of Movements*. My aim is to contribute to the preservation of the Alexander Technique as I have been taught it through my research and practice and in this the opportunity to work on Bernstein continues to prove an inspiration and a privilege.

Birmingham, UK   Rose Whyman

Acknowledgements

Special thanks to Dr. Victoria Door for her help, Dr. Martin Leach and friends and colleagues in the Professional Association of Alexander Teachers www.paat.org.uk. Thanks also to the Whyman, Crabtree and Waldron family for all your encouragement. The task of translating this book was made much easier with the help of Tanya Lipatova from the University of Birmingham, UK, with the first six chapters. Dr. Vera Talis, Senior Researcher at the Institute for Information Transmission Problems (Kharkevich Institute) RAS in Moscow, helped with the whole translation and contributed the excellent introductory essay. Thanks to Professor Mark Latash of Penn State University, USA, for his support. Dr. Roland Ennos at the University of Hull, UK, and Dr. Michele Grimm at Michigan State University, USA, made many useful suggestions and corrections. The work of Dr. Irina Sirotkina and Dr. Roger Smith is inspirational and I thank them for their support of all aspects of my work. Thanks also to Jayne Smith and Lizzie Compton of the University of Birmingham. Finally, my thanks go to Merry Stuber, Editor for Cell Biology and Biomedical Engineering at Springer, for her encouragement and enthusiasm for this project, as well as to Murugesan Tamilselvan, Anthony Dunlap and all the team at Springer.

Introduction

Nikolai Bernstein (1896–1966) was a Russian scientist, the author of the theory of construction of movements in humans and animals. According to this theory, coordination of movement is organised hierarchically, with many levels, and motor learning is a process of transitioning a specific movement to the control of the appropriate level. Bernstein, who was a physician, a specialist in mechanics and a mathematician, proposed an entirely original definition of the coordination of movement as the central nervous system's process of overcoming the redundant number of degrees of freedom of the movement apparatus in humans. The theses he developed about the role of the central command and sensory corrections through the reflex circle have become the basis of contemporary motor control theory. (As a young contemporary of Ivan Petrovich Pavlov (1849–1936), Bernstein criticized the application of Pavlov's reflex theory in a simplified way to living movement, where muscle contraction is initiated by the reflex arc from external stimuli. Elaborating the 'physiology of activity' throughout his life, Bernstein showed that any living motion is initiated from within the organism through the reflex circle, where the central neural system can get feedback about movement realization.)

Up until now, two of Bernstein's books have been translated into English. The first, *The Co-ordination and Regulation of Movements*, was edited by Bernstein himself and published in 1967, a year after his death.[1] The second book, *O Lovkosti i ee Razvitii* (*On Dexterity and its Development*), is a popular interpretation of his major book, *O Postroenii Dvizhenii* (*On the Construction of Movements*), for which Bernstein was awarded the Stalin Prize for science in 1948. In 1949, soon after finishing *On Dexterity and Its Development,* Bernstein began to be persecuted. For this reason the book was not published during his lifetime. A proof copy was discovered after Bernstein's death by his student and first one of his biographers I.M. Feigenberg and was published in 1991. In 1996, *On Dexterity and Its Development* was translated into English and published with extensive commentaries

---

[1] Bernstein, NA (1967) *The Co-ordination and Regulation of Movements.* Oxford, Pergamon Press.

by the distinguished American scientists Mark Latash and Michael Turvey.[2] From the centenary of Bernstein's birth in 1996 to 2000, many papers by Bernstein and his co-authors were translated and published with commentaries in the journal *Motor Control*. Before you now is the translation of Bernstein's third book, *Biomechanics for Instructors*, which is a popular version of his book of 1926, *Obshchaya Biomekhanika* (*General Biomechanics*).

Biomechanics for Instructors is a clear, seemingly simple description of the structure of the human body as a machine. It is an entertaining story, presented from an evolutionary point of view, about how the muscles work, how the skeleton is constructed and how the joints move.

The book is unique in that we have the opportunity to 'hear' live dialogue between Bernstein and students, his pedagogic technique and his way of conveying the difficult questions of mechanics. (Only a few two-minute long documentary films of Bernstein, broadcast in 1948, which were shown before the main feature in cinemas are known of today.) *Biomechanics for Instructors,* which has been republished several times in Russia,[3] began to interest the translator, Rose Whyman, although her speciality is far removed from biomechanics and the physiology of movement. In studying stage movement and how the Alexander Technique can facilitate this, Rose could see answers in Bernstein's book to her practical questions about the structure of the human body, the working of its muscles and how the joints are stabilised. She could also see the book as a popular interpretation of complex questions of the control by the brain of the human movement apparatus.

It must be emphasized that translating Bernstein's texts is complicated because of vividness of his speech and his rather creative word formations. He sometimes uses clauses that last the length of a whole paragraph (!), and correspondingly, there is a great concentration of thought within them. *Biomechanics for Instructors* is written in a relatively easy style, in the form of a dialogue with the auditorium, with speech that is at times simple, colloquial and every day, including archaisms, and the translator has dealt splendidly with this, preserving the very spirit of the times.

The mid-1920s in Russia was a time when a mass of young, poorly educated people poured into industrial and cultural life as a result of the changes to society after the October Revolution of 1917. It was also a time when a great number of educated and well-off people left Russia and it became a time of challenge for those remaining. Nikolai Bernstein's father, Aleksander Nikolaevich Bernstein, the well-known Moscow psychiatrist, a student of the famous S. S. Korsakov,[4] founded *Tsentral'nyi Preemnii Pokoi* (The Central Hospital Ward). This was the first clinic in Moscow where mentally ill people were kept without restraints. Aleksander's

---

[2] Latash, ML, Turvey, MT (Eds.) (1996) *Dexterity and Its Development*. Mahwah, NJ, LEA Publishers.

[3] Bernstein, NA (2001) Biomekhanika dlya Instruktorov (Biomechanics for Instructors) In Shestakov MP (Ed.) *Izbrannie trudi po biomekhanike i kibernetike* (*Selected works on biomechanics and kybernetics*) Moscow, SportAcademPress. [In Russian].

[4] Sergei Sergeevich Korsakov (1854–1900) was one of the greatest neuro-psychiatrists of the 19th century, author of works in neuropathology, psychiatry and forensic medicine.

work was inspired by the psychiatry of his time, where in both Europe and Russia experimental scientific approaches to diagnostics were being developed and at the same time that there was progress in treating mentally ill people more humanely.[5]

After the Revolution, Bernstein senior became the deputy head of the Main Administrative Board of the Scientific Institutions of the Academic Center of Narkompros.[6] He worked in the Commission for Reform of Medical Education and in 1920 founded, and became director of, the Moscow State Psychoneurological Institute.

The close relationship and the influence of his father on Nikolai Bernstein is evident in their correspondence of 1914, when the 18 (!) year old son, entering the historical-philological (!) faculty of Moscow State University, wrote to his father 'I found an essay about Gogol[7] in Korolenko,[8] where he expresses the same proposition about his cyclothymia[9] as you and refers to Bazhenov's 1902 essay, which you perhaps know.[10] He points out that Gogol's father suffered from the specific features of the very illness that you have described; the alternating periods of feverish activity, the inventiveness, which coincides with manifestations of his comic talent and other times of dark depression, apathy and a horror of death, from which, it would appear, he died'.[11] The characteristics of this letter—punctiliousness in citation, a particular feeling for words and a psychological aspect to the scientific approach—are remarkable in Nikolai Bernstein's work throughout his life. It is a great sadness that these qualities of the brilliant scientist were later used against him, when at the end of the 1940s in Russia he was stigmatised for 'cosmopolitanism' because of his citations of foreign authors and for 'idealism' at the expense of 'materialism' because of the figurativeness of his comparisons. It is surprising but the ideas expressed in the letter from 1914 quoted above were ideas that persisted through

---

[5] See Gilyarovsky, VA (1922) Pamyati A.N. Bernsteina (A.N. Bernstein Obituary). *Zhurnal Psykhologii, Nevrologii i Psykhiatrii* (Journal of Psychology, Neurology and Psychiatry) 1, pp. 5–9. [In Russian].

[6] *Narodnyi Kommissariat Prosveshchenie* (The People's Commissariat for Education), *Narkompros*, was the Soviet agency for administration of public education and culture.

[7] Nikolai Vasilievich Gogol (1809–1852) was the great Russian writer famous for works such as the novel *Myortvye Dushi (Dead Souls)* and the play *Revizor (The Government Inspector)*.

[8] Vladimir Galaktionovich Korolenko (1853–1921) was a Russian writer, journalist and human rights activist. The essay referred to is Korolenko, V.G. (1909) 'Tragediya velokogo yumorista: Nieskol'ko myslei o Gogol'e' ('The tragedy of great humanist: several thoughts about Gogol'). *Russkoye Bogatstvo* (Russian heritage), 4, 6. p. 27. [In Russian].

[9] Cyclothymia is now referred to as bipolar disorder.

[10] Nikolai Nikolaevich Bazhenov (1857–1923) was a Russian psychiatrist, who worked for the humane treatment of mentally ill patients. In 1916 he emigrated first to France then returned to Russia from Belgium in 1923 when he was very ill. The essay referred to is Bazhenov, NN (1902) Bolezn' i smert' Gogolya' (The Illness and Death of Gogol). Public reading at the annual meeting of The Moscow Society of Neuropathologists and Psychiatrists. Kushnerev@K, b. 1, p. 38. [In Russian].

[11] Talis, VL (2019) *Doktor kotoryi lyubil parovoziki (The Physician Who Loved Steam Engines)*. Manuscript. [In Russian].

Bernstein's life, as one of his final notes in 1965 in the popular journal *Science and Life* is entitled 'Death from the fear of death'.[12]

After Bernstein entered the historical-philological faculty of Moscow State University, he transferred to the medical faculty, moved by a patriotic desire to take part in the First World War, which was just beginning. After he graduated from the University he served as an army doctor in the Red Army from December 1919 to March 1921. After demobilisation, he came to the Psychoneurological Institute where he headed the laboratory until 1922 and published his first work in 1922.[13] This publication appeared in the first issue of the journal founded by his father and was an obituary to A.N. Bernstein who died aged 52 in May 1922, at the peak of his creative energies. In August of the same year N.A. Bernstein had moved to the famous *Tsentralnyi Institut Truda* (TsIT), Central Institute of Labour, established on the personal instruction of V.I. Lenin.[14] In summer 1924, Bernstein was already presenting, as a member of the TsIT delegation at the First International Congress on the Scientific Organisation of Labour in Prague, the results of registering the movement of humans by his own method of cyclogrammetry (see Lecture 8 in this book). In 1926, he wrote his first book, *General Biomechanics*, dedicated to the memory of his father. The founder and leader of TsIT was Aleksei Kapitonovich Gastev, a metal worker from the Russian city of Suzdal, a revolutionary, poet and later on a victim of Stalin's repressions, shot in 1939. We gain an insight into the atmosphere of those years from his poem:

LET US ERECT A MONUMENT
TO THE AMOEBA—the creator of reaction,
TO THE DOG—our greatest friend,
called to exercise.
TO THE MONKEY—the hurricane
of living movement,
TO THE HAND—the wonderful
intuition of will
and construction,
TO THE BARBARIAN with his stone
strike,
TO THE INSTRUMENT,
as the banner of will,
TO THE MACHINE—teacher
of accuracy and speed
and TO ALL BRAVE HEARTS,
calling for the REMAKING OF HUMAN BEING.

---

[12] Bernstein, NA (1965) Smert' ot stracha smerti [The death from the fear of death]. *Nauka i Zhizn'* (*Science and Life*) 2, p.149 (1965). [In Russian].

[13] Bernstein, NA (1922) Kvoprosu o vospriyatii velichin (o roli pokazatel'noi funktsii v prozessakh vospriyatiya velichin) (On the question of the perception of quantities (about the role of exponential function in the processes of the perception of qualities)). *Zhurnal Psykhologii, Nevrologii i Psykhiatrii* (J of Psychology, Neurology and Psychiatry) 1, p. 21–54. [In Russian, abstract in German].

[14] Vladimir Ilyich Ulyanov (1870–1924), better known as Lenin, the head of the government of Soviet Russia from 1917 to 1922 and of the Soviet Union from 1922 to 1924.

CURSED be all
cowards,
hypocrites,
obscurantists,
those who howl and shriek
on the roads
and in marketplaces
where our machine races along.
HELLO!
Hello to our happily
dashing our
IRON,
FULL-BLOODED,
SURE
MONTAGE![15]

Biomechanics for Instructors came out in 1926 after N.A. Bernstein left TsIT, seemingly because of the incompatibility of the 'romantic' proletarian approach of Gastev with the academic approach of Bernstein, a third generation physician and scientist. His sceptical attitude to TsIT is evident at the beginning of Lecture 10 in this book, where we read: 'the research that I carried out on labour movements in TsIT ceased after my resignation....' TsIT, which was set up to be self-financing, not only carried out research but also trained workers. In Lecture 10 a member of the student audience notes, 'The lads who have trained at TsIT were also told to chop within a single plane but then you do not get the force. If you take it out from the body, it is somehow easier'. Including this wonderful example of 'simple-hearted' speech (most likely having invented it himself), Bernstein begins to compare the 'vertical' strike in chiselling, which was taught at TsIT and the strike 'to the side'. In recounting the example of the strike 'to the side', which gives the most force to the strike and which was later on to be a classic of the physiology of movement, the 29-year-old lecturer exclaims rhetorically:

Is it possible then that all vertical strikes are so hopelessly bad? Regrettably this is so. Is it possible then that all the strikes to the side are irreproachably good? Well, no, it is impossible to say that. I shall reveal a little secret to you. I have shown you all kinds of vertical strikes and I have chosen only the best of those to the side that came out as a result of detailed investigations.

In this way, having revealed a little of the principles of experimental work of the time which still apply in the present day, Bernstein displays his unique ability to see the most characteristic features of living movement visually without the statistics in use today. In fact Fig. 6 in this book becomes the first experimental figure in his main book, *On the Construction of Movements* (1947), illustrating his idea that:

cases when, during a specific movement, very different muscle groups are activated at different times and in different patterns, as compared to what could be expected from elementary anatomical analysis, are much more common than cases when muscle action is classical

---

[15] Gastev AK (1973). Trudoviye ustanovki (Organization of labour) in Gastev Yu.A, Petrov E.A. (Eds.) *Trudoviye Ustanovki*, Moscow, Economika. [In Russian].

and fully understandable. There are many movement elements, where we are still unable to interpret the behavior of each muscle group; sometimes, analysis of external and reactive forces can be performed and it clearly reveals the logic of those unexpected at the first glance muscle actions, but this logic is very much different from the elementary logic of high-school textbooks.[16]

Lecture 6 is dedicated to efficient stance in labour movements, and here the audience pose themselves the question of 'which angle between the feet is more correct in filing and chopping—67 or 70 degrees?' The answer is:

Perhaps the extent to which all these values and proposals are random and groundless is not as evident to you as it is to me, from my perspective. These values are often just spun out of thin air to give some firm rule and expound how scientific one's approach to the question is.

After this astonishing observation (which is still relevant today!), Bernstein, having defined the motion of the hammer in the strike, evaluates the corresponding movement of the common centre of gravity of a worker with a hammer in their hands and, having clarified that, concludes for effect that 'Obviously, all the debates about 67 and 70 degrees completely fall away'.

It must be noted that as in *Lectures on Biomechanics for Instructors*, all Bernstein's further experimental and even theoretical work has a clear practical application, whether it is the calculation of moments of force in running with highly professional sports people or the force of strikes on the piano. In this way 'the investigation and improvement of the working place of the tram driver' of 1928 becomes in 1934 'the project of the efficient work place of the metro driver'.[17] On leaving TsIT in January 1925, Bernstein began a study of the biodynamics of strikes on the piano in the Gosurdarsvennyi Institut Muzykalnykh Nauk (The State Institute for Musical Sciences), which has the evocative acronym 'GIMN'.[18] Among the colleagues who went with him from TsIT to GIMN was Tat'yana Sergeevna Popova (1902–1992), the wife of his younger brother, Sergey.[19] Tat'yana Sergeevna, who was educated in the mathematical faculty of Moscow State University, came from a millionaire merchant family of clothing manufacturers—the Popovs. She, like another member of this family, the famous avant-garde painter Lyubov Sergeevna Popova (1889–1924), was without doubt one of the most outstanding figures of her time. In 1924, in the period of work at TsIT, she writes:

The Central Labour Institute is a new institute…Everything is done in a new manner, not in the way it was done by the bourgeoisie. The Institute is striving to introduce science into production. The interests of the director are those of a metalworker, therefore the Institute

[16] Bernstein NA (1947). *O Postroenii Dvizhenii (On the Construction of Movements)*, Moscow, Medgiz, p.24. [In Russian].

[17] Autobiography of NA Bernstein of 29.05.1945 from the personal files of NA Bernstein in the Archive of Medical Science, Moscow (F. 9120. V. 8/3. Unit 19).

[18] The acronym GIMN spells the word for 'hymn' in Russian.

[19] Sergei Alexandrovich Bernstein (1901–1958) was a professor of structural mechanics, author of fundamental calculation and trials of engineering constructions.

studies mostly the work of a metalworker and his two main procedures: chiselling and filing.[20]

The study of movements of musicians is an area that still has been studied little in spite of the great interest in it,[21] because of the methodological and experimental complexities of the registration and interpretation of the subtle highly coordinated movements involved. Bernstein entered this area as a musician with absolute pitch, having composed music and played professionally on the piano and oboe. According to legend in the Bernstein family, Rachmaninov prophesied a career in music for him.[22] In the essay 'Research on the Biodynamics of the Piano Strike' of 1930, Bernstein and Popova write that 'most studies in the area of piano methodology originate either from physiologists who are dilettantes in music or from musicians who know nothing about physiology'. This essay, translated in 2003, is introduced by the following words; 'Bernstein and Popova report an impressive study (even by modern standards) of a complex motor behaviour: movements performed by concert pianists, specially a series of octave strikes made with one hand'.[23]

In the years 1927 to 1933, the work of Bernstein and his colleagues was often published in German in the journal *Arbeitsphysiologie (Labour Physiology)*. From the 1930s, the work of Bernstein and his colleagues was in the main devoted to the biodynamics of locomotion—walking, running, jumping with a run up and so on. It is interesting that the investigation of locomotion began initially at the behest of Narodnyi Kommissariat Putei Soobshcheniya (The People's Commissariat for Railways), and the first paper was published in 1926 in the collected articles of *Voprosy Dynamiki Mostov (Questions of the Dynamics of Bridges)*.[24] Later Bernstein's group studied 'normal walking, loaded walking, changes in walking due to tiredness and in the restoration period, age-related changes to walking (children's walking development, elderly gait), gait in amputees with prosthetics, central-nervous pathology of walking, the running of outstanding sportsmen, the development of running in children, the technique of the long jump with a run up etc.'[25] In mid-September 1929, Nikolai Bernstein took abroad a description of human walking

[20] Talis VL (2015) 'New pages in the biography of Nikolai Alexandrovich Bernstein' in Nadin, M. (Ed.) *Anticipation: Learning from the Past. The Russian/Soviet Contribution to the Science of Anticipation.* Cham, Springer, p. 313–328.

[21] See, for example, Altmüller, E, Wiesendanger, M., Kesselring, J. (Eds.) (2006) *Music, Motor Control and the Brain.* Oxford, Oxford University Press.

[22] Rachmaninov, Sergei Vasilevich (1873–1943) was a Russian composer and virtuoso pianist. Following the Russian Revolution, Rachmaninov and his family left Russia; in 1918, they settled in the United States.

[23] Kay, BA, Turvey, MT, Meijer, OG (2003). 'An Early Oscillator Model: Studies on the Biodynamics of the Piano Strike (Bernstein and Popova, 1930)', *Motor Control*, 7, p. 1–24.

[24] Bernstein NA (1927) 'Issledovaniya po biodynamike khod'bi i bega' (Studies on Biodynamics of Walking and Running) in *Sbornik: Voprosy Dynamiki Mostov, trudi Narkomata Putei Soobsheniya, (Collected Works: Questions of the Dynamics of Bridges. Works of The People's Commissariat for Railways)* Vol.63, p.51–76.

[25] Autobiography of N.A. Bernstein of 29.05.1945 from the personal files of N.A. Bernstein in the Archive of the Russian Academy of Medical Science, Moscow (F. 9120. V. 8/3. Unit 19).

in the form of an atlas, hoping that Springer would publish it. When the plans for publication fell apart he left *The Atlas of Human Walking* as a gift for the Dortmund Institute of Labour before his departure to Moscow at the beginning of January 1930.[26]

Bernstein's trip to France and Germany (for about 3 months at the end of 1929) was dedicated, on the one hand, to the buying and ordering in Europe of experimental equipment for the Moscow laboratories and, on the other, to instructing his European colleagues in his technique of recording movements, which was advanced for the time. The head of the Russian physiological school, an academician from a Russian aristocratic family, Alexei Alexeevich Ukhtomsky writes:

> A young Russian scientist N.A. Bernstein showed the striking example of how Fischer's method can be used for the complete mechanical assessment of various movements in the process of physical labor ... Having registered the trajectory of a hammer in the process of manual labour, Bernstein calculated vectors of acceleration for different points of the trajectory and for different positions of the moving centers of gravity. Knowing kinetic energy in different sections of the trajectory makes it possible to figure out where the kinetic energy of the movement of the particular system in the direction of the trajectory will be the highest. Obviously, this will be the most favorable moment for the technical application of this particular manual labour movement. Besides the technical significance of such analysis of labour movements, it also presents an inspiring interest from the purely scientific point of view. Not one existing method of registering the motor reactions of organisms provides the extent of completeness and objectivity that cyclogrammographic method does. And not one of the existing methods of researching motor reactions possesses the visual representation and precision of the cyclogrammometric method. No doubt this method has a tremendous future.[27]

Later on, in 1933, in an essay dedicated to the development of physiology during 15 years of Soviet power, A.A. Ukhtomsky talks about:

> the works of Professor N.A. Bernstein that are excellent both in design and completion ... here we will experience a new revolution in natural sciences, the consequences of which we are currently not able to realize just like the contemporaries of Leeuwenhoek and Malpighi were not able to foresee the changes that the invention of the microscope would bring to the future generations.[28]

Bernstein had collected a great deal of experimental material about the movements of healthy subjects and patients. He constantly questioned how the brain directs the 'mechanical machine' called the 'human body' (see Lecture 1). In calculating joint torques, he separated those that are the result of the joints' interaction from those that are brought about by the influence of the force of gravity and those active muscle forces that are put into action by the motor nerves from the central nervous system. He analysed the complexity of the mechanics of this system, where the muscles, which carry out movement, can change their mechanical properties themselves depending on the conditions of the work being executed (this is discussed

---

[26] A reprint of this Atlas was recently published (2019) http://d-nb.info/1193129338

[27] Cited from E. Loosch & V. Talis (Eds) (2014). Feigenberg I.M. *Nikolai Bernstein. From Reflex to the Model of the Future*. Berlin: LIT, p.44–45.

[28] Ibid., p. 47–8. Antonie van Leeuwenhoek (1632–1723) was a Dutch scientist and Marcello Malpighi (1628–1694) an Italian biologist and physician.

in Lecture 2). As he had no knowledge yet about muscle spindles—the sensory endings of the muscles were discovered by C.S. Sherrington in 1929—Bernstein presupposed their existence in the feedback circle of the motor apparatus in humans and animals.[29]

Later Bernstein formulated the so-called 'model of the desired future' based on the probability prognosis of the movement being realised. This concept then became the basis of his 'physiology of activity'. In 1935, Bernstein's seminal paper 'The Problems of the Interaction of Co-ordination and Localization' was published. In this, Bernstein discussed the 'principle of equal simplicity', which is illustrated by the fact that everyone, independently from the spatial metric of movement, writes his/her signature with the hand or the foot identically in any scale, or draws a circle with the hand extended in front or to the side, in spite of the fact that the muscles participating in this movement are completely different. Later on, this paper, which became the forerunner of the cybernetic approach, became Bernstein's gift to the 'father of cybernetics', Norbert Wiener, when they met each other in Moscow in 1960.[30]

A wave of anti-Semitism travelled through the USSR at the end of the 1940s, under the banner of 'the struggle with cosmopolitanism'. This reached the area of physical culture and sport, in which Bernstein was mostly working at that time, and he fell victim to it. The nature of this persecution, like the witch hunts of the middle ages, is shown, for example, in the protocols of the party and social meetings of collectives where Bernstein worked.[31] In these meetings, on instruction from the highest powers in the Soviet Union there are critiques of sports scientists connected with 'one anti-party group of theatrical critics'![32] In March 1949, the editorial in the newspaper *Sovetskii Sport* entitled 'Let us finally unmask cosmopolites and their yes-men' inveighed against Bernstein.[33] At the same time Communist Party Bureau sessions and meetings of scientific councils of the Institutes of Physical Culture in Moscow and Leningrad and the Institute of Prostheses in Moscow, with which Bernstein was then collaborating, issued a directive which threatened to Bernstein's work situation. In 1950, there was a critique of Bernstein in the newspaper *Pravda*, the central organ of the communist press in the USSR, in the essay 'Against the vulgarisation of the theory of physical education'.[34] *Soviet Sport* rushed to echo this essay in September 1950 with 'Where are the textbooks for physical education insti-

---

[29] Sir Charles Scott Sherrington (1857–1952) was an English neurophysiologist, histologist, bacteriologist, and a pathologist, Nobel laureate and president of the Royal Society in the early 1920s.

[30] Chkhaidze, L.V., Chumakov, S.V. (1972) 'Formula Shaga' (Formula of the Step) in *Fizkultura i Sport* (*Physical Culture and Sport*). Moscow, p. 90. [In Russian].

[31] Personal files of N.A. Bernstein in the Archive of the Russian Academy of Medical Science, Moscow (F. 9120. V. 8/3. Unit 19).

[32] Personal files of D.D. Donskoi in the Central State Archive of The Institute of Physical Culture (F.9, № 1, Unit 260–292)

[33] *Sovetskii Sport* (*Soviet Sport*), 29, 22 March (1949). p.1. [In Russian].

[34] Zhukov P, Kozhin A. (1950) *Pravda* (Truth), 21 August. p.233. [In Russian].

tutes and technical schools?'[35] In spite of this, in 1950, the presidium of the Academy of Medical Sciences asked the scientist to start work in the Institute of Neurosurgery, where in six months, from January to June 1951, Bernstein organized a laboratory for the physiology of movement and by November was already conducting experiments. But all these efforts were in vain—in January 1952 he was transferred with his equipment to the Institute of Neurology. On 19 March 1953, the laboratory of physiology and the pathology of movement there was liquidated. Soon after, Bernstein applied to leave work on health grounds. He did so and then had no work. His wife was disabled due to asthma and they had an adolescent son, so he earned money from translations for journals of abstracts. For instance, 'in the period from I.IV.1955 to I.V.1959 725 abstracts are completed...published in the journals *Biology* and *Mathematics'*.[36] Heavy daily work gave Bernstein modest means for living and the opportunity to follow the progress of physiological science abroad.

A new wave of interest in Nikolai Bernstein began among young scientists in connection with the so-called thaw which began at the end of the 1950s in the USSR and in Russian science, in connection with the interest in the new science called 'cybernetics' developed by Wiener, that investigates the control principles in technical and biological systems. In 1957, Bernstein was invited to make a presentation at the Department of Mechanics and Mathematics of Moscow State University in the seminar of Alexei Andreevich Lyapunov (1911–1973) and then in the Moscow physiological seminar organised in 1958 by Israel Moiseevich Gelfand (1913–2009) and Michael Lvovitch Tsetlin (1924–1966). His advanced mathematical qualification allowed Bernstein on the one hand to publish strictly applied works on the mathematics about frequency analysis in biological data that is for instance in the electroencephalogram[37] and, on the other, to develop the theoretical bases of modelling in biology with the mathematician I.M. Gelfand and the physicist M.L. Tsetlin. Bernstein was famous throughout Moscow and far beyond as a scientist who had retired from work, but who could be consulted by both physicians and musicians on questions of both mathematics and medicine. Every half hour new visitors with their experimental data came to his door on Bolshoi Levshinskii Lane in Moscow, where at this time he was living with his wife and son in only two rooms. Nikolai Alexandrovich valued the data highly, but he had no opportunity to run experiments and obtain this himself. The pensioner Nikolai Aleksandrovich Bernstein acted as an opponent for dissertations, wrote forewords for books, and invited young people to write joint articles for encyclopaedias.[38]

---

[35] Bloch L, Kosvinzev B, Nikolaev A. (1950). *Sovetskii Sport* (*Soviet Sport*). 104. 2 September. [In Russian].

[36] Personal files of N.A. Bernstein in the Archive of the Russian Academy of Medical Science, Moscow (F. 9120. V. 8/3. Unit. 19).

[37] Bernstein, N.A. (1962). K analizu neperiodicheskich kolebatel'nykh sum s peremenennymi spectrami po metody vzveshenny khreshetok (Towards an analysis of aperiodic oscillatory sum with the variable spectra using the methods of weighted lattices). *Biofizika* (*Biophysics*), 7,4 p. 376–381. [In Russian].

[38] See, for example, Bernstein N, Kotz J (1963) 'Tonus'. *Bolshaya Meditsinskaya Entsyklopedia* (*Great Medical Encyclopedia*), 2nd edn. 32. p. 418–422. [In Russian].

In 1959, when the first cosmonauts were being trained, Bernstein advised about the possibilities for the coordination of movements of humans in zero gravity. In 1965 and 1966 Bernstein's keynote papers 'On the Roads to the Biology of Activity' and 'The Immediate Tasks of Neurophysiology in the Light of the Modern Theory in Biological Activity' and the popular article 'From the reflex to the model of the future' were published. In them, Bernstein considered the concept of 'activity' in its general biological meaning as the reason for the development and evolution of living beings. He warned about the unjustified use of mathematics in biology, stating 'do not attempt to concretise too soon the electrophysiological intracerebral picture of the phenomena by means of externally observed relations. In these last works, as before, accuracy in citation is evident, so also is the literary figurativeness of Bernstein's text and the genius's foresight in seeing the unfounded attraction with modelling of those scientists whom I.M. Gelfand calls 'modellers' and V.S. Gurfinkel calls 'keyers' (from the word 'key', having in mind those people who have one key for everything and stick it in without discrimination wherever it can be stuck!).

On the 16th of January 1966, N.A. Bernstein died from liver disease, in the house where he lived all his life on Bolshoi Levshinskii Lane, just after publishing his book *Outline of the Physiology of Movements and Physiology of Action*[39] but just missing seeing the publication of his first book in English, '*The Coordination and Regulation of Movements*'. Knowing, as a doctor, that liver disease is incurable, he discharged himself from the polyclinics in the last months of his life and, fearing to upset his family, left a note saying that when he passed out they should call for Volodya, a student he was close to, the future neuro-rehabilitation specialist Vladimir Naidin (1933–2010).[40]

Today, reading *Biomechanics for Instructors*, we see distinctly in the one part of the auditorium the young lecturer, talented, highly educated in literature, medicine, mathematics and engineering and in the other the students from whom 'The most I can ask… is knowledge of the four rules of arithmetic and basic knowledge of mechanics and technical drawing' (see the Foreword). For almost a hundred years since his book came out the science of movements has developed considerably not only due to development of technology, electronics, etc. but also to a great degree due to approaches developed by N.A. Bernstein. Equipment in a contemporary biomechanics laboratory will include a system for registering movements; a force platform to register the deviation of the centre of mass of the subject standing on it, which in Bernstein's time had to be calculated with complex methods; and electromyographic apparatus to register the activity of the muscles and sometimes, depending on scientific goals, a tomograph, electroencephalograph, etc. Bernstein only had at his disposal a system for registering movements that was groundbreaking for his time but less precise than contemporary ones. However, it must be admitted that the

---

[39] Bernstein, NA. (1966) *Ocherki po fiziologiii dvizhenii i physiologii activnosti. (Outline of the Physiology of Movements and the Physiology of Activity)* Moscow, Medizina.

[40] Talis V.L (2019). *Doktor kotoryi lyubil parovoziki. (The Physician Who Loved Steam Engines).* [In Russian].

high level of accuracy, which is essential, for example, for the registration of movements of the pupils of the eyes, is not necessary for movements as slow as the movements of a healthy person and even a sportsperson, so this does not diminish the value of Bernstein's book.

Thus, for example, in Lecture 9 Bernstein describes the graphic method of defining speed and acceleration of body segments—a task which contemporary PC programmes calculate analytically very quickly. Then, Bernstein, knowing the mass of the segment, graphically finds the force developed by this segment when it moves and concludes:

> All the calculations that have been described might seem to you to be laborious and boring. On the other hand, how fascinatingly interesting it is when out of a lifeless cyclogram, which is like a motley net of points, suddenly one by one all the secrets of the movement which has taken place begin to appear!... You begin to feel that you have learned to read some language that was incomprehensible to you before.

Of course, what in Bernstein's time '...took about 1 month to analyse...with contemporary techniques... can be done in seconds'[41] but this may not always help depth of understanding of the questions being researched. The contemporary reader might not accept the material of this lecture as guidance for action but all the same will get an unique sense of how the picture of the kinematics of movement grows into the picture of forces producing this movement and can gain an impression therefore of the so-called problem of inverse dynamics.[42] Today in the laboratories studying living movements, the quality of force analysis of human movements depends on the elaboration of models of the human body, which, as before, are far from ideal in their approximation to living movement.

The complexity of this modelling today is still connected with the kinematic particularities of living joints, as discussed in Lecture 1. Explaining the notion of 'degrees of freedom', Bernstein says, 'Let's see now what the mobility of bones which are joined together is, in relation to one another and how to define this mobility. I am warning you that this is a rather complicated question'. It echoes a contemporary textbook on biomechanics where we read 'Degrees of freedom represent the kinematic complexity of a biomechanical model'.[43] What then happens when such difficult material is explained? Bernstein asks a volunteer from the audience to rotate his arm in the shoulder joint, then to flex and extend the elbow, explaining that in the first case we are dealing with two degrees of freedom and in the second, with one degree of freedom. The contemporary textbook continues, 'The degrees of freedom (dof) correspond to the number of kinematic measurements needed to completely describe the position of an object'. This is, of course, an exact definition, developed by the analytical approach, but it does not minimize the importance of the mnemonic approach of Bernstein, which he uses throughout *Biomechanics for*

[41] Latash ML, Zatiorsky VM. (2016) *Biomechanics and Motor Control.* New York: Academic Press, Elsevier.

[42] Latash ML (2008). *Synergy.* Oxford, Oxford University Press.

[43] Knudson D. (2007) *Fundamentals of Biomechanics,* 2nd Edition, New York, Springer.

Instructors. Bernstein's extensive description of various joints, including the particularities of the cartilage in them, appears to be a great supplement to contemporary textbooks, which are illustrated by splendid drawings, instant photographs and even references to video materials!

N.A. Bernstein is acknowledged today as the 'father' of motor control and not one textbook or book on biomechanics, kinesiology and sport physiology gets by without citing him. The hierarchical principles of movement control that he proposed are used in the construction of robots.[44] Music teachers learn the motor learning principles proposed by Bernstein. However, few are capable of being guided by them in teaching children music. Music teacher V.A. Guterman, who died in 1993, is a happy exception to this. The well-known Soviet physiologist L.A. Orbeli (1882–1958) said about her method of teaching music that as a result of her authorial 'method of instruction, based on the system of control of the teacher's kinetic and tactile sensations juxtaposed with the kinetic perceptions of the pupil, the latter learns by subtle analysis of these sensations, by correct evaluation and the ability to control his muscles, creating the necessary co-ordinations'. Describing her method, V.A. Guterman writes 'Genuine, great musicians…have each created their own individual technique, growing out "of their model of the desired future" (according to N.A. Bernstein), from their ideals for the sounds'.[45]

Guterman's method is concordant with E.V. Maximova's original method of 'abilitation' of people with autism spectrum disorder. She, a biologist by education, and well acquainted with Bernstein's theory of construction of movements, noticed in her patients 'loss of entire control levels according to Bernstein's classification'. In her book she writes, 'The children can walk, run, comb their hair, but cannot make any reaching movement (or these movements are very difficult for the child'. Having noticed this, Maximova came to the conclusion that 'the levels of construction of movements proposed by N.A. Bernstein can be regarded more widely, as levels of psychic reaction'. Maximova's method is based on the assumption that 'in many psychic pathologies disruption of perception is primary and disruption of communication, disruption of emotional reaction, disruption of behaviour, are secondary'.[46] Making use of methods of bodily therapy, Maximova tries to build up step-by-step the 'lost' control levels according to Bernstein's classification beginning with 'tonic', the 'more ancient' and finishing with the most complex, such as the highest cortical level of speech and writing, the most difficult for the patients in this group.

---

[44] Poramate Manoonpong, Geng, T, Kulvicius, T, Porr, B, & Wörgötter, F (2007). 'Adaptive, Fast Walking in a Bipedal Robot under Neuronal Control and Learning'. *PLOS Computational Biology.* Vol. 3 p. 1–16.

[45] Guterman V.A. (1994). *Vozvrascheniye k Tvorcheskoi Zhizni. Professional'niye zabolevaniya ruk.* (*Return to the creative life. Professional diseases of the hands*). Ekaterinburg Gumanitarno-ecologickii Litsei. [In Russian].

[46] Maximova E.V. (2008) *Urovni obscheniya. Prichiny vozniknoveniya rannego detskogo autisma i ego korrekstsia na osnove teorii N.A. Bernshteina* (*Levels of communication. Reasons for the emergence of autism spectrum disorder in early childhood*). Dialogmifi: Moscow. [In Russian].

It must be noted that today, although the levels structure of movement control is apparent, concrete morphological substrates of localisation of one or another level remain undefined as they were in Bernstein's time, in spite of the progress of brain research. But all the same, why do not only specialists in motor control but also educationalists, trainers, musicians, physiotherapists and mathematicians find in Bernstein through his works a teacher and ally? The answer is very simple: through his genius, the combination of his deep insight in several branches of science led to the birth of the new science of movement control, of which he is considered the father. The outstanding mathematician of our time I.M. Gelfand said at the funeral of physiologist N.A. Bernstein that we are burying 'an outstanding mathematician'. Despite this, we can say that Bernstein is still not fully evaluated and understood by his compatriots (the essay about Bernstein in the German Wikipedia is 2.5 times longer than the Russian one).

Bernstein has left us two popular books, one of which was translated 20 years ago and the second lies before you now! Like Bernstein's students, listening to Bernstein in the cold Moscow of 1925, let us arm ourselves 'with the four rules of arithmetic and basic knowledge of mechanics and technical drawing' and go forward keeping in mind the logic of the outstanding scientist with a tragic fate— Nikolai Aleksandrovich Bernstein!

Vera L. Talis

Contents

Foreword 1

Lecture 1 3

Lecture 2 17

Lecture 3 29

Lecture 4 45

Lecture 5 59

Lecture 6 75

Lecture 7 93

Lecture 8 109

Lecture 9 127

Lecture 10 141

List of Figures

Fig. 1.1 The human bony skeleton 4
Fig. 1.2 Diagram showing the development of the limbs of vertebrates
At the top—the limbs of the lizard, in the middle—the limbs
of a four-legged mammal. At the bottom, the mechanism
of pronation and supination (see Lecture 3) (by Braus) 7
Fig. 1.3 A longitudinal section of three vertebrae On the left—the porous
vertebral bodies that are joined with cartilage cushions;
on the right—the spinous processes of the vertebrae
that are linked between themselves by ligaments.
In the middle—there is a canal for the spinal cord with
openings for nerves to enter and exit (by Spalteholz) 8
Fig. 1.4 Elbow joint of the right arm, shown from the front.
The pulley of the humero-ulnar joint and the ball
of the humero-radial joint are seen clearly. A—humerus,
Б—ulna and B—radius (by Tol'dt) 11
Fig. 1.5 A section of the ball and socket hip joint (by Mollier) 11
Fig. 1.6 Saddle joint between bones A and Б, allowing
two degrees of mobility 13
Fig. 1.7 Diagram of a longitudinal section of the knee joint.
One can see the concavo-concavo cartilage lining (by Mollier) 14
Fig. 1.8 Cross section of the bones of the lower leg. On the left
the fibula and on the right the tibia (by Spalteholz). 15
Fig. 1.9 Positioning of bone partitions in the head of the femur
(on the right), in comparison with the lines of stress
in a crane (on the left) 16

Fig. 2.1   The action of muscle on the bone lever. This muscle
            (biceps or the two-headed muscle of the upper arm)
            acts only on the forearm bones, so that squeezing
            the wrist into a fist, depicted here, is a whim of the artist 19
Fig. 2.2   Stretching of a spring with a weight. The white arrows
            pointing upwards represent the tension of the spring;
            the arrows pointing down represent the force of gravity
            of the load; the black arrows represent the balancing
            force of both forces. x_1 is the point of equilibrium
            (the resultant force = 0) 21
Fig. 2.3   This is how the waning oscillations of the spring look,
            disturbing the balance of the weight. The waning
            oscillation of the muscle (Fig. 2.4) looks very similar
            but it dies away quicker there. 22
Fig. 2.4   (A) The curve of the muscle action current, (Б) the curve
            of the twitch of the muscle, (B) the oscillation of the tuning
            fork which marks hundredths of a second.
            This is a photographic record (by Yudin). 25
Fig. 2.5   (I) Composition of two consecutive single contractions. (II)
            Composition of many contractions following on from
            one another. (III) Tetanic contraction (by Landois) 26

Fig. 3.1   The right shoulder blade from behind and the types
            of its mobility. 1—adduction, 2—abduction, 3—elevation,
            4—depression, 5—rotation inwards, 6—rotation outwards 30
Fig. 3.2   An X-ray of the shoulder joint. On the right is seen the
            humerus with its ball-like head and on the left we see the
            scapula with its hollow. At the top we can also see
            the shoulder blade's process for the collar bone (by R. Fikk) 31
Fig. 3.3   How the turns of the shoulder blade (A) influence
            the borders of mobility of the shoulder (Б) (by Mollier). 32
Fig. 3.4   General limits of mobility of the shoulder (by Mollier) 34
Fig. 3.5   Model which shows the way the radius (Б) is fixed to
            the ulna (Б) and the humerus (Б) (by Braus). 36
Fig. 3.6   The joint between a vertebra and the ribs. The arrow
            shows the axis, around which the rib can rock (by Mollier) 38
Fig. 3.7   The right hip joint from behind, with all its tight-fitting
            muscles removed so that one can see the ligament,
            which forms a spiral (by Spalteholz) 41
Fig. 3.8   The arch made by the bones of the foot and its muscle
            and its tendinous-muscular tension brace (by Mollier) 41
Fig. 3.9   Abridged plan of the body. See explanations in the text 43

Fig. 4.1 Muscles of the neck on the right side. The muscle
          which inclines the head (sternocleidomastoid muscle)
          extends obliquely along the whole length of the drawing
          (by Spalteholtz) 47
Fig. 4.2 These two (diagrammatically depicted) muscles are
          completely equal to each other 48
Fig. 4.3 Section of an insect joint and the positioning of muscles
          that operate this joint (by Shenikhen) 52
Fig. 4.4 Sketch of muscle tension braces of a one-axis joint
          (muscle antagonists). 54
Fig. 4.5 Sketch of the arrangement of muscle tension braces
          of a two-axis joint 54
Fig. 4.6 Muscles of the right shoulder region, from the front.
          (A) Pectoralis major. (Б) Deltoid (by Spalteholz) 57
Fig. 5.1 Scheme of the muscular tension braces of the four-legged
          mammal. A—ventral tension brace, Б—tendinous tension
          brace of the spine, B—spinal tension brace. 60
Fig. 5.2 Front part of the muscle bandage which fixes shoulder
          blade to the trunk. Serratus anterior (by Spalteholz) 62
Fig. 5.3 Muscles of the back. M = trapezius; Б = latissimus dorsi
          (see Fig. 5.6); Б = deltoid (by Spalteholz). 63
Fig. 5.4 Direction of the muscle pulls of the short group of the shoulder:
          (1) Supraspinatus, (2) infraspinatus, (3) subscapularis,
          (4) coracobrachialis, (5) pectoralis minor 65
Fig. 5.5 The directions of muscle pulls of the long group of shoulder
          muscles. (1, 2, 3)—separate bundles of deltoid,
          (4)—pectoralis major and (5)—latissimus dorsi 67
Fig. 5.6 The position of latissimus dorsi on both sides and the way
          they attach to the humeri. All other back muscles are
          removed for clarity (by Mollier) 68
Fig. 5.7 Muscles that are responsible for the movement
          of the wrist (hand). Left side—view of the right forearm
          from the front. Right side—the same view from the back.
          (1)—radial flexor. (2)—ulnar flexor (3–4)—radial extensor.
          (5)—ulnar extensor (by Mollier) 71
Fig. 5.8 The muscles which control the movements of the fingers.
          (I)—the extensor of the fingers, (II)—the deep flexor,
          (III)—the superficial flexor (by Mollier) 73
Fig. 6.1 View of the right shoulder blade from behind 76
Fig. 6.2 Diagram of the organization of the muscles of the front
          and hind limbs of a four-legged mammal. The corresponding
          muscles are depicted by the same points and indicated by the
          same numbers. 77

Fig. 6.3   Flexors (C) and extensors (P) of the hip joint and their action
            (by Mollier) 77
Fig. 6.4   The muscles which raise the thigh outwards (ПН)
            and the muscles "adducting" the thigh (ПР) (by Mollier).......... 78
Fig. 6.5   The two-headed flexor of the knee (A) and the extensor
            of the knee (B) (by Mollier)................................. 80
Fig. 6.6   The calf muscle and its role in standing (by Mollier) 81
Fig. 6.7   Diagram of a skeleton, showing the positions of the centers
            of gravity. Г—center of gravity of the head. T—center
            of gravity of the torso. П—center of gravity of the upper arm.
            ПR—center of gravity of the forearm. Б—center of gravity
            of the upper leg. Гол—center of gravity of the lower leg.
            ЦТ—center of gravity of the whole body (by R. Fikks) 87
Fig. 6.8   Center of gravity of two joined segments (S) and the
            method of finding it. The center of gravity of A lies at a_1,
            and the center of gravity of segment B at a_2. The explanation
            of the rest of the letters and lines is in the text................. 88
Fig. 6.9   Hinged person for determining the positions of the centers
            of gravity. Explanation in the text (by O. Fischer)............... 89
Fig. 6.10   Moving the center of gravity, S, of two points, A and B,
            of which one is ten times heavier than the other.
            Explanation in the text.................................... 90
Fig. 6.11   The arrangement of the feet in normal standing. The circle
            is the projection of the center of gravity of the body. The area
            of support of the body is shaded in. The side of each
            square is 5 cm.. 91

Fig. 7.1   Nerve fibres under a microscope. The conducting part is
            in the middle surrounded by the isolating cover (by Rozenthal)..... 94
Fig. 7.2   The diagram of the reflex arc. On the left is a cross section of
            the spinal cord, on the right the nerve paths. The bold dotted
            line is the centripetal (sensory) neuron. The bold continuous
            line is the centrifugal (motor) neuron 99
Fig. 7.3   Cross section of half a brain, somewhat schematised.
            (A) The outgoing bundle of motor nerve fibres,
            (Б) the incoming bundle of sensory nerves—(B),
            the nerve cells (by Deiters) 100
Fig. 7.4   View of the human brain from the left side. Л—the frontal
            part of the left hemisphere. B—the temporal part
            of the left hemisphere. 3—the occipital part of the left hemisphere.
            T—the parietal part of the left hemisphere. Ц—the central
            part of the left hemisphere (where the motor centres are situated).
            P—the fissure of Rolando (now generally referred to as the central
            sulcus or central fissure). M—the cerebellum. П—the medulla
            (by Flatau) 101

Fig. 7.5   A section of the brain with the front at the top and the back
           below (in the direction of a peaked cap band). Above there
           are the motor centres, in the middle the sensory centres
           and below the corresponding centres of the old brain
           (by Flatau) 103
Fig. 7.6   A diagram of the conducting paths of the old brain.
           ЗБ—optic thalamus; П—globus pallidus; M—cerebellum.
           The triangles are the intermediate nuclei. The black triangle
           and the circle below depict the centres of the spinal cord with
           the centripetal and centrifugal nerves coming out from them 104

Fig. 8.1   The cerebral cortex under a microscope. The black spots
           are the nerve cells. On the right are individual cells that
           have been magnified (by Campbell) 110
Fig. 8.2   The conducting paths of the new brain. KП—cerebral cortex;
           the other designations are as in Fig. 8.6. ПП—pyramidal path 111
Fig. 8.3   The sensory and motor centres in the cerebral cortex of the brain . . 112
Fig. 8.4   Snapshot of a pole vault. Which of you readers has seen
           by eye such a position of the body? (from Lorentz photo
           catalogue). 115
Fig. 8.5   Photographs of a horse running, taken by Muybridge. 117
Fig. 8.6   The fall of a cat, which had been held up by its legs,
           and how it turns while in flight (from Anschütz' photographs). 117
Fig. 8.7   Photographic apparatus for cyclical snapshots and an
           electrical motor with rotating shutter (installation by the
           author at TsIT) 119
Fig. 8.8   Cyclogram of walking by Marey. The subject is walking
           from left to right. The movement of the head, right hand
           and right leg has been photographed (by O. Fischer) 120
Fig. 8.9   A cyclogram of walking by Braune and Fischer. The subject
           is walking from the left to the right. The point at the top is
           the crown of the head and the stripes from the top down are
           the upper arm, lower arm, hip, lower leg and foot; the square
           in the middle of the snapshot is the scale (by O. Fischer) 121
Fig. 8.10 A cine series of the swing strike of a blacksmith's hammer.
           The upper line is the stroke and the recoil, the second and
           third lines are the swinging movement, the lower line is
           the striking movement (by Fremon). 122
Fig. 8.11 The same swing strike as in the previous picture but taken
           on one plate with the use of the rotating shutter. The path of
           the movement of the sledgehammer is very distinctly visible
           (by Fremon) 122
Fig. 8.12 A subject with small lamps marking his joints 124

Fig. 8.13   Half of the stereoscopic cyclogram of the chisel chopping
             movement. The strike is near to normal 8 (see Lecture 10).
             Taken by the author at TsIT   126
Fig. 8.14   The laboratory arrangement of the cyclographic recording
             (set up by the author at TsIT). A—Distributing electrical table
             where the direction of all the recordings takes place.
             Б—Photographic camera. B—Rotating shutter with four slots.
             Г—Light of 1000 candle power. D—Centimetre scale.
             E—Workbench for the subject   126

Fig. 9.1   A cyclogram of walking photographed by the author at TsIT.
             Only the hip, knee and ankle joints of the right leg have been
             recorded. The direction of walking is from left to right.
             The walk is a ceremonial march   128
Fig. 9.2   A cyclogram of chiselling, recorded without a rotating
             shutter. The vertical stroke (see Lecture 10). Photographed by
             the author at TsIT   128
Fig. 9.3   A section of cine film, showing chiselling in the form of
             an animation scheme. The film was prepared from
             the cyclograms   131
Fig. 9.4   Graph of the component velocities of speeds of movement
             of the hammer in chiselling. _____ velocity of movement
             forward and back; - - - velocity of movement up and
             down; . . . . velocity of movement right and left   134
Fig. 9.5   A cyclogram of filing, taken using a camera with a sliding
             cassette. From left to right the traces are (1) the right elbow,
             (2) the right wrist, (3) the right index finger and (4) the left
             thumb. Taken by the author at TsIT   137
Fig. 9.6   Diagram of a set-up for photographing adapted
             by the author for experiments with the spatial sense at TsIT.
             A—Photo camera, Б—A subject, B—observer   138
Fig. 9.7   (a) A photograph of walking blindfolded in a triangle,
             taken from a height. What is visible: the triangle chalked
             on the floor and the trace of the small lamp on the top
             of the head of the person being tested. The arrow indicates
             the direction of movement. The task was executed well.
             (b) The same experiment but with another person being
             tested, whose ill condition is well reflected in the photo.
             The light dots above are the person being tested and the
             researcher, taken from a 'bird's-eye view'. Both photos
             were taken by the author with Dr. N. Ozeretskii   139

Fig. 10.1 The consecutive positions of the right hand with the
          hammer during chiselling. The long axes of the upper arm,
          forearm and hand are conventionally depicted with straight
          line links. The white links and the hammers are the
          swing-up, and the black—the strike. The difference in
          time between the adjacent positions is 1/15th of a second.
          The strike is near to normal 8 142
Fig. 10.2 The consecutive positions of the arm with the hammer
          during incorrect (vertical) strike. The designations are the
          same as in Fig. 10.1 143
Fig. 10.3 The speeds of movement of the parts of the right arm with
          the hammer during the correct strike (chiselling): _____ speed
          of the centre of gravity of the hammer; ------- speed of the fingers;
          .... speed of the wrist; ·—·—· speed of the elbow. Further
          below—tenths of seconds; the divisions on the left—metres
          per second; from 0 to 0.4 s is the swing-up, from 0.4 to
          0.6 the strike.. 145
Fig. 10.4 Cyclogram of a strike to the side, close to Normal 6.
          Photographed by the author at TsIT........................ 146
Fig. 10.5 Cyclogram of the vertical strike. Photographed by the
          author at TsIT... 147
Fig. 10.6 Cyclogram of chiselling. The vertical strike with a big recoil.
          Photographed by the author at TsIT........................ 147
Fig. 10.7 A cyclogram of chiselling. The vertical strike with an
          excessively deep raising of the hammer. Photographed by
          the author at TsIT....................................... 148
Fig. 10.8 A cyclogram of chiselling. The strike approaches normal 4.
          Photographed by the author at TsIT........................ 150
Fig. 10.9 The raising of the arm and the hammer according to
          normal No. 4, 6 and 8. View 1 from behind and 2 from
          the right 152

Foreword

This course consists of lectures that I delivered in July and August 1925 on courses of further training for instructors of industrial education from *Moskprofobr*.[1] It is a reworking of shorthand reports of the actual course, but it has been fundamentally reworked, almost every passage has been rewritten. Live speech to an audience is always very different from the forms of speech that we are used to seeing in print; it is, whether you like it or not, fragmentary, transient and not always strictly planned. The expressions of interest or surprise or questioning on the faces of the audience, which are not seen by a person reading shorthand reports, always influence lecturers, making them digress, make their expressions more exact and explain parts that are not understood.

Nevertheless, the form of live speech has been very dear to me; I remain convinced that conveying information in such a way stimulates the readers' initiative and makes them actively follow the developing chain of thought. I carefully preserved all the questions, comments and even the mistakes of my listeners, because they are natural and normal and readers will undoubtedly put themselves into the place of my interlocutors from the audience.

I just completed the first volume of a large guide to biomechanics entitled *General Biomechanics,*[2] which is now being published by R. I. O. VTsSPS in summer this year.[3] Naturally, many sections of this new book became part of the course

---

[1] *Moskprofobr*: Moskovskii podotdel professional'no-tekhnicheskogo obrazovanie (Moscow subdivision of professional technical training). This institution oversaw the varied professional and technical schools and training after the 1917 Revolution.

[2] Bernstein's *Obshchaya Biomekhanika (General Biomechanics)*, published in 1926, was his contribution to the developing science of biomechanics, including his foundational work on human locomotion. Though Bernstein was the first scientist in the world to develop the study of movement as a way of studying the workings of the brain, this book has never been translated into English.

[3] Redaktsionno-izdatel'skii otdel Vsesoyuznogo tsentral'nogo soveta professional'nykh soyuzov. Editorial and publishing section of the All-Union Central Council of Trade Unions.

© Springer Nature Switzerland AG 2020
N. A. Bernstein, *Biomechanics for Instructors*,
https://doi.org/10.1007/978-3-030-36163-1_1

that was delivered and this reworking of it, although in a very different form. However, it follows that there is the question of whether it was worth publishing this short course in these circumstances and whether it is a repetition of what has already been said somewhere else.

No, this is by no means a repetition. To begin with, despite everything I tried to do to make the book *General Biomechanics* easily accessible, it is not easy for young people whose knowledge is at a beginner's level to understand.[4]

As it is the first manual on this subject in Russian, it has to be detailed, and so I could not avoid using Latin names and mathematical expressions. The common content between this course and the book is rewritten in a significantly popularised and shortened way and, I think, could be accessible in this form to anyone who has finished trade school[5] or factory-and-works school.[6]

The second and main justification for this course is that it also differs significantly from the book in plan, content, and in character of exposition. Two of the lectures in the course are devoted to the introduction of methods of studying movements and a short history of this study that are not touched upon in *General Biomechanics*. The last lecture is devoted to an analysis of typical labour movements—work with a hammer—which, again, I am publishing for the first time in a popular form.

I did not aim to give this course a strictly applied character, to make it into something like a collection of recipes. It is applied only to the extent that it gives a short systematic analysis of the human machine and its potentialities; its task is to teach the reader to think mechanically in relation to a student—trade school students or a factory-and-works school students—in the same way as they think about a machine. If this task is fulfilled at least to some small degree, then the practical aim of the course will be achieved.

The course is designed for a reader who has had little preparation. The most I can ask from the reader is knowledge of the four rules of arithmetic and basic knowledge of mechanics and technical drawing. The future will show whether this course can suffice for independent study of biomechanics. In the meantime, I intend it to be an aid to revision for those who have already got, in some form or other, education in biomechanics from a physical culture group, factory-and-works school or from my lectures etcetera. Therefore, this course is published *as a manuscript*.

I would like to express my gratitude to comrade M.V. Pshenichnikov, who took on the idea of publishing this course with enthusiasm and did so much to ensure that this happened. I warmly invite all those comrades who heard this course in person to send me their objections, doubts and questions.

---

[4] Bernstein writes 'first grade' and in 1923–4 this referred to the education of children aged 8–13.

[5] *Proftekhshkola* (Trade school) Profsoyuznaya shkola dlya izuchenie profsoyuznoi raboty.

[6] *Fabzavucha* (Factory-and-works school) Fabrichno-zavodskoye uchenichestvo dlya podrostkov pri predpriyatii. FZU schools offered manufacturing apprenticeships for young people in industry and were trade schools offering basic professional education or vocational training.

Lecture 1

In this lecture, Bernstein introduces the science of biomechanics with reference to the bony skeleton, joints and degrees of freedom of mobility.

Comrades! In an exact translation, 'biomechanics' means 'the mechanics of life'. In essence, this is the science of how the living machine, that is, each of us, is constructed, of how the moving parts of this machine are arranged and how they work. Knowledge of the living machine is necessary so that, by skilful use of it, the best and most productive work can be achieved.

You understand that the laws of mechanics are the same everywhere, whether in relation to a steam engine, a machine tool or the human machine. This means that we do not have to develop new, special laws of mechanics. We only have to formulate the description and the characteristics of the living machine as we would for a car, a loom, etc.

The difference between the two forms of description is only that the human machine is much more complicated and intricate than any artificial machine that there has ever been. Therefore, there are far more gaps in our knowledge of biomechanics than, for example, in knowledge of the mechanics of construction or the science of machines. Here, there is still a great deal that is obscure, about which we can only speak approximately.

The other difference is this. Each of us can hope to invent some new artificial machine or device, then get a patent for it and immediately transform the idea into action, that is, construct one's invention. Regrettably, to perfect the human machine, to subordinate its construction to one's will is impossible at the moment. We have to train it as it is, with all its virtues and shortcomings. Only indirect, circuitous methods are available to us in order to bypass its shortcomings and to make full use of its advantages.

Let us not waste time and turn to the investigation of the living machine, its arrangement and ways of using it.

The human body consists of a series of segments, united between themselves by hinge joints and capable of rotating in relation to one another. We can consider each

© Springer Nature Switzerland AG 2020
N. A. Bernstein, *Biomechanics for Instructors*,
https://doi.org/10.1007/978-3-030-36163-1_2

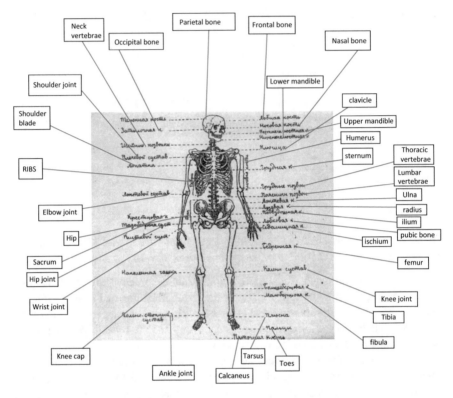

Fig. 1.1 The human bony skeleton

of these segments initially as firm, with an unchanging form. For example, each of your fingers consists of three segments. It is the *bones* that serve as a support for these segments, but we are not really interested in a detailed description of them. For biomechanics, the question of how the segments of the human machine are joined onto each other and what their mobility relative to one another is like is much more important.

The human skeleton (Fig. 1.1) consists, rounded up, of 170 separate bones, which are all more or less mobile relative to one another. However, for us to consider all the joints that they have between them would be going too far. First of all, we are going to simplify substantially the skeleton of the human machine, turning it into what we will term the *abridged plan* of the human body.

Later, we will take the whole torso with the neck as one whole and will no longer pay attention to the mobility of its parts relative to one another, which is in fact quite significant. We will consider the head as another such unchanging segment, leaving aside its internal mobility (the movement of the lower jaw, etc.). We will divide each of the extremities into three parts. Then our abridged plan will look like this[1]:

---

[1] 'The Russian word *plecho* means the shoulder and upper arm. Bernstein also uses it as a shorthand for *plechevaya kost*', arm-bone (humerus).

```
                            Head
                             1.
                              o
                 2.           s          2.
              Upper arm       r          Upper arm(see footnote 1)
                 3.           o
                              T
          Lower arm                       Lower arm
                          5.     5.
           4.             h      h             4.
          Hand            g      g            Hand
                          l      i
                          h      h
                          T      T
                          6.     6.
                          g      g
                          e      e
                          l      l

                          r      r
                          e      e
                          w      w
                          o      o
                          L      L

       Foot 7.                          7. Foot
```

We designate the joints between these segments as follows:

1. Cervico-occipital joint
2. Shoulder joint
3. Elbow joint
4. Wrist joint
5. Hip joint
6. Knee joint
7. Ankle joint

You must be wondering why the human body came to have an appearance like that in the diagram. It is only when you have got to know how the skeleton of vertebrates came about and how it developed you can appreciate the reasons for this. After all, there wasn't an engineer taking part in its creation who could present us with the project of its construction and an explanatory note of justification.

Let's begin with the torso. If you look at the construction of the torso of a fish or snake, you will see that it is a chain of separate small segments, linked to each other. From each segment—the vertebra—on each side to bony little cross-beams—the ribs, linked by elastic tension braces—the muscles. This is the original plan of construction of the vertebral animal. The human foetus is constructed exactly in this way at the beginning of its development. A little flexible stick—the spinous string— is the supporting rod of its body at the beginning; next, in its place, the chain of bony vertebrae—the vertebral column—develops. Muscle tension braces arise from all sides of the spinous string, which strengthen the spinous string and at the same time ensure its flexibility.

The *extremities* are a newer addition to this ancient system. In all the vertebral animals, fish, reptiles, birds and mammals, the extremities are constructed according to one basic plan: in essence they have not changed.

The basic diagram of the extremities is like that of the *hand*: it is united to the body by means of one bone; with this one bone are joined two, etc., until in the end a limb like this turns into a whole bunch of bones, which are arranged like rays. Fish fins, for example, are constructed in such a way.

The limbs of human beings are constructed in a quite similar way. The top segment of the extremity (upper arm, thigh) consists of one bone; the next segment (forearm, lower leg) consists of two bones; next there are small bones in several rows, and in the end the final segments (hand, foot) five rays—fingers and toes.

Since our limbs stopped carrying out the duties of fins and got accustomed to moving the body on dry land, they have undergone a series of changes. We can find the original form of limbs adapted for walking in lizards (Fig. 1.2, top). All four limbs of the lizard are situated at a right angle towards the spinal column, so that the thigh and shoulder bones lie horizontally and the lower legs and forearms are directed downwards. Limbs have little ability to do work when they are constructed like that. First of all, because they have moved far apart to the sides they cannot support the weight of the animal's torso and the lizard has to drag it along the ground (to crawl). Secondly, its limbs are not adapted for swinging backward and forward, as is necessary for walking. In order to walk, the lizard has to lean on one leg and then with the help of the muscles of the whole torso has to turn around this leg in a circle, like a pair of compasses. It is easy to understand how uneconomical this is.

The limbs of mammals went through curious changes in comparison with this plan. Imagine that the hind limbs have turned forward by 90 degrees and the front legs turned backwards by the same amount, so that we get the structure that is depicted in the middle diagram of Fig. 1.2. Here, too, the top part of limb consists of one bone, the middle of two and the lowest of many small bones, but their position relative to the torso is already completely different.

First, it will be clear to you that in this new situation, it is now much easier to support the weight of the torso, because the supports are situated right under the torso and not far away from it at the sides, as before. Also, the middle joints, which are slightly bent towards each other, serve as kinds of springs: they can pool their resources and soften the shocks coming from behind as well as from the front.

Secondly, as you will understand, as the axes of all joints of the limbs have now turned in the transverse direction, the limbs can now swing forward and backward freely; that is, they can make exactly the movements that are necessary for walking. It is apparent that walking can now be carried out more easily and with less effort than it was in reptiles.

We have left out one event that inevitably accompanied the turning of the limbs that I have just described. You see that the front legs turned at a right angle *backwards*—that means that their lower segments (hands) had to turn backwards

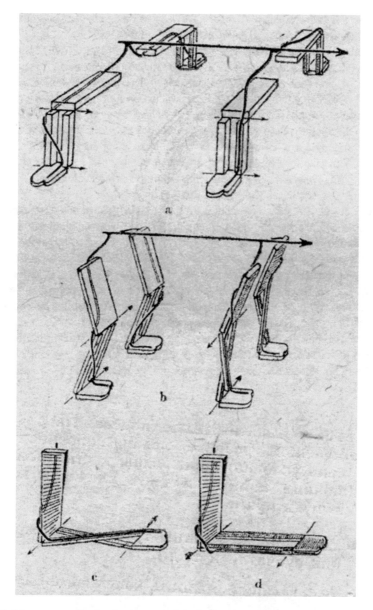

Fig. 1.2 Diagram showing the development of the limbs of vertebrates
At the top—the limbs of the lizard, in the middle—the limbs of a four-legged mammal. At the bottom, the mechanism of pronation and supination (see Lecture 3) (by Braus)

as well. Indeed, in order to put them in the right position, with fingers forward, the middle part had to turn around its axis so that the bones had to cross like the letter x. Such a turn exists in mammals; a human can do this but it is not permanent: a human can do this turning at will. We will come back to this movement.

There is no significant difference between the structure of the human body and the four-legged mammals, apart from some change in proportions. Nevertheless, a human being has to stand vertically and walk on two legs, so the mechanical conditions of the work of his/her body are completely different from those of the four-legged animal. Meanwhile, the whole structure of the human skeleton is adapted to the conditions of a four-legged way of life. There are still a huge number of remnants in it that still have not been adapted for two-legged existence. Many details of the human machine that would be completely understandable if we walked on four legs turn out to be absolutely unsuitable and even harmful for the two-legged way of life. It is sufficient to mention the structure of the female pelvis; everybody knows how much more difficult and painful and dangerous childbirth is for a human being than it is for four-legged creatures. The need for maternity clinics depends in great part on the fact that people started walking vertically without changing their motor constructions.

Let's have a look at the structure of joints.

The bones that are joined together are not directly in contact with each other; they are covered with elastic articular cartilage in the place where the joint is. The cartilages of both bones fit each other quite exactly; if one bone has a head at the end then the second one ends with a hollow exactly of the appropriate size and form. The whole joint is enclosed in an impenetrable capsule, inside of which there is a small cavity in this way. This capsule strengthens the joint and at the same time serves it as a lubricating apparatus; the internal membrane secretes drop by drop a liquid, which constantly lubricates the surfaces of the cartilages as they rub against each other.

The type of joint that has just been described is the predominant type in the human body. It is the most perfect construction, but the human being can have cruder, ancient forms. A longitudinal section of three vertebrae is depicted in Fig. 1.3. You can see

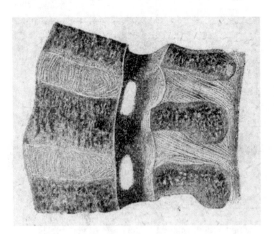

Fig. 1.3 A longitudinal section of three vertebrae
On the left—the porous vertebral bodies that are joined with cartilage cushions; on the right—the spinous processes of the vertebrae that are linked between themselves by ligaments. In the middle—there is a canal for the spinal cord with openings for nerves to enter and exit (by Spalteholz)[2]

---

[2] Werner Spalteholz 1861–1940 was a German anatomist, author of a book published in English as *Hand Atlas of Human Anatomy*, 1901–3.

that the way they are joined is very simple: a flexible cartilage cushion is situated between them, which provides vertebrae with some reciprocal mobility. Between such stiff cartilage joints and a joint of the present day there are various transitional forms, which we are not going to look at.

The main difference between human joints and mechanical joints is in the way that parts are fastened to each other. Mechanical bearings for the most part are constructed so that one part entirely encloses the other, so that a hard link is made between them. A human being doesn't have such enclosing devices and therefore the strengthening link is made in a different way. You will remember that the whole joint is covered with an impenetrable capsule. This capsule consists of quite firm tendinous tissue, which is for the most part strengthened further by auxiliary ligaments. Thus the capsule itself can provide some firmness to the joint. But this is not all. Imagine that we are trying to pull the bones that are joined together away from each other. But because this joint capsule is airtight, you will be making an attempt to increase the size of the joint cavity, without letting air in it. The air pressure will resist this process. It will be as difficult to move the joint bones apart as it would be to move apart two hollow hemispheres when they have been put together and the air has been pumped out. The strength of the air pressure in the hip joint comes to 1½ pood[3]; that is, it is twice as much as the weight of the whole lower limb. Scientists carried out the following experiment. They removed all the muscles from the lower limb of a corpse making it hang on the joint capsule only. They also hung a weight on this limb from below. The limb kept firmly in its place; but they only had to make a little hole in the capsule for the air to come noisily into the cavity straightaway and the joint surfaces immediately came apart.

That is not all. The joints are surrounded with muscles from all sides. These muscles are attached to both bones of the joint. Muscles for their part have an ability to stay firm when they are stretched. Besides, they are constantly somewhat stretched and facilitate strengthening the joint even more than the air pressure.

Let's see now what the mobility of bones that are joined together in relation to one another is and how to define this mobility. I am warning you that this is a rather complicated question. Could one of our comrades come out here and show how he can move his upper arm segment in the shoulder joint? You can see that he can rotate the shoulder forward and backward and to the sides. You know that a rod that is fixed at one end, but with the other end free to move in all directions, will constantly be able to move this end along the surface of a sphere. The lower end of the shoulder segment moves along a similar spherical surface. Moreover, as its mobility is not limited to just one line but by the whole surface area, we can say that it has two degrees of mobility. That's not all. I will retain the lower end of the shoulder segment of our volunteer, to keep it immobile, in one position. Can he move his shoulder in any other way in such a situation?

---

[3] A *pood* is Russian measure of weight equivalent to 16.38 kilos.

Students: No, he cannot.

Lecturer: Is that so? Bend your arm at the elbow at a right angle. Can you move your arm as if the forearm was a spoke and the upper arm represents the axis that is turning round itself?

Volunteer: I can.

Lecturer: However, the lower end of your shoulder which I am holding with my fingers doesn't change its position in space. Therefore the shoulder has, in all, three degrees of mobility. Thus, we speak about one degree of mobility when the segment can move only along one line, about two degrees when one end can move along a whole surface and about three degrees when the part can also turn around its longitudinal axis. Now try yourselves, and tell me how many degrees of mobility does your elbow joint have?

Students: One degree.

Lecturer: Absolutely correct; this is flexing and extending the elbow: you see that the lower end of the forearm can move only along the arc of a circle. The human body has also joints with two degrees of mobility. You see for example the ankle joint. Keep your lower legs immobile and try to do all the movements you can with your feet. You can see that the toes can move in various directions along some surface but the feet are not able to rotate along their longitudinal axis.

Now let's ask ourselves the question of what forms should the joint ends have in order to give the various degrees of mobility. Let's start with one degree of mobility.

It is best if we consider mechanical joints first. It is clear that a simple cylindrical bearing has only one degree of mobility. Indeed, if the axis of a wheel is connected to a machine tool with the help of such a bearing then every point of the wheel can move only along one single line—the circumference. That means that the cylindrical type of joint will have only one degree of mobility. But if our cylinder doesn't have flanges, then it can move along its axis as well; that is, it would reveal a second degree of mobility. Gears in a car gearbox behave in such a way, for example. However, if the cylinder has flanges, then it is clear that the form of these flanges can be of any kind, but the body on the whole has to be round and should be what we call a *rotary body*.

Such things as pulleys, spools, round samovars, columns, etc. can be called rotary bodies. All joints that have rotary bodies as their parts will therefore have one degree of mobility.

Let's consider, as an example, joints between the humerus and the ulna. You can see (Fig. 1.4) that the humerus ends with a spool at the lower end. On the end of the ulna, which is joined to the humerus, there is a hollow of corresponding shape, which is restricted both from the top and from the bottom by two bony protrusions. If the spool and this groove fit close together then they will be mobile relative to one another, in one direction. That means that any body that is tightly joined to the spool will also have only one degree of mobility in that particular joint. We have already had the chance to see that the elbow joint is indeed a one-degree joint.

Fig. 1.4 Elbow joint of
the right arm, shown from
the front. The pulley of the
humero-ulnar joint and the
ball of the humero-radial
joint are seen clearly. A—
humerus, Б—ulna and
B—radius (by Tol'dt)[4]

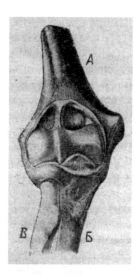

Fig. 1.5 A section of the
ball and socket hip joint
(by Mollier)[5]

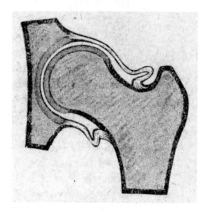

We can imagine other types of one-degree joints as well. For example, such is the connection between the piston cross head and its guides in a steam engine, the connection between a screw and a nut, etc.. However, this type of one-degree joint is not seen in the living machine. Let's turn to three-degree joints now. After what we have discussed you will easily understand that a connection by means of a ball head has three degrees of mobility. All three-degree joints in the human body are made using exactly this type of ball joint.[6] Figure 1.5 shows a section of the hip joint of a human body. You can see that the hip ends with the perfect ball head at the top and that this

---

[4] Carl Toldt (1840–1920) was an Austrian anatomist and author of *An atlas of human anatomy for students and physicians*, translated into English in 1919.

[5] Siegfried Mollier (1866–1954) was the author of *Plastische Anatomie: Die Konstruktive Form des Menschlichen Korpers* published in 1924.

[6] Bernstein's note: 'The three degree joint between the jaws is far more complicated and is not considered to be a ball joint'.

head fits tightly into the cavity of the hip bone, which has the form of a hemisphere. If you check it on yourself, you will easily be convinced that the hip has three degrees of mobility in relation to the body in exactly the same way as the shoulder does.

There are no surfaces in nature that could give two and only two degrees of mobility in relation to each other if they fit one another tightly at the same time.[7]

Whichever surface you choose it will either give you one degree of mobility in relation to a similar surface or three degrees straightaway in all (or maybe none, if one surface cannot be moved from another without having broken their contact).

Let's check with an example. Let's take two surfaces that touch each other, for example, this slide, which is lying on the table. How many degrees of mobility does it have in relation to the table?

Students:	Two degrees.
Lecturer:	Which two degrees?
Students:	It can move along the surface.
Lecturer:	Do you want to compare this movement with the movement of the end of the humerus along a spherical surface? In this you would be right, but after all, this slide can also rotate around itself, like the upper arm, on every point of the table surface where you move it, thanks to the two degrees of its mobility.
	This will be the third degree of mobility. Besides, a plane is after all only a particular aspect of a sphere: it can be regarded as part of a sphere of infinitely large cross section. Let's have a look at other surfaces. Suggest some yourselves.
Students:	What mobility does the wheel on a rail have?
Lecturer:	If the wheel can only roll along the rail but cannot slide then its degree of mobility is one. However, this example is not suitable because the wheel does not touch the rail with the whole surface. Give another example.
Student:	A stopper in a carafe?
Lecturer:	What do you think?
Student:	One degree?
Lecturer:	If the stopper has the form of a cone then of course it is one degree— rotating around its axis. When you take the stopper out of the carafe you immediately break its contact with it. And can't you find an example of a surface that has no degree of mobility in relation to another surface area? Is that difficult for you? An example is a printing plate and the matrix from which it is made. Could you, if you put one onto the other move them without breaking their contact? Clearly, you could not. In the human body there are joints exactly of this type, with no degree of mobility. The joints of the bones of the skull are an example of this.

However, let's go back to the mobile joints of the living machine. I have already mentioned that it has two-degree mobility joints and they are not cylindrical. How can we reconcile that with the new deduced rule about the mobility of surfaces?

---

[7] Bernstein's note: 'The only exception, which was mentioned before, is a cylinder'.

Fig. 1.6 Saddle joint between bones A and Б, allowing two degrees of mobility

The fact is that joint surfaces in the human body are somewhat pliant and flexible. Therefore, they can keep mutual contact even when they do not exactly fit each other. The living machine knows several types of joints that become possible only thanks to the flexibility of the cartilages. Saddle and ovoid joints are of this type, for example.

Imagine an area that has a form of an English saddle, i.e. which is convex in one direction and concave in the other. If another surface of similar form and without any flexibility in this position is in contact with this surface, then they won't have any mobility in relation to one another. But if they can change their form a bit then the mobility will be a two-degree one. If we took a saddle as one of the surfaces, let's take a rider as the second surface. Indeed the rider can, without changing the position of his/her legs, move forward and backward and can slide from one side of the saddle to the other. But he/she cannot rotate his/her body around its axis, that is, turn right or left. The joint between the carpal and the metacarpal bone of the thumb, has, for example, such a saddle-like construction (Fig. 1.6).

Another type of two-degree joint is also possible only thanks to the pliancy of the cartilage and is called an ovoid joint. Imagine a cavity cut in a section from the side of the flatter part of an eggshell. If you try to insert a whole egg into the cavity from the side then it will also have two degrees of mobility in it: you will be able to rock it in all directions but won't be able to turn it around the vertical axis in the cavity like you would a spinning top. The joint between the head and the first cervical vertebra belongs to this type.

In the human body, there are joints that make use of the flexibility of cartilage even to a greater extent. In Fig. 1.7, you can see a longitudinal section of the knee joint. You can see that the surfaces of both connected bones do not fit each other; to begin with, both of them are convex. In order to provide a wide contact between two such dissimilar surfaces, a third intermediate cartilage which has a concavo-concavo form is attached between the two bones (thigh bone and tibia). Thanks to this, the joint as a whole achieves either one or two degrees of mobility, depending on its position and the degree of pliancy of the joint tendons of an individual.

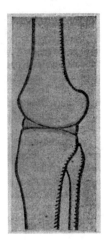

Everything that we have been talking about up until now is only in relation to the method or character of mobility of the joined parts. Nothing here has yet been determined about the *limits of movement* in the given joint. It's clear that one-degree and three-degree joints can both be highly mobile and have very little mobility. For example, the cartilage union of two vertebrae has at least three degrees of mobility because the springy cartilage is equally pliant in all possible directions. But the vertebrae are mobile to a very small extent in each of these directions, by only 5–10 degrees. On the other hand, the elbow joint with only one degree of mobility has very wide limits of mobility: 140 degrees and more. We will consider the limits of mobility of separate main joints of the human body and the ways of describing these limits at a later stage.

So far, we have looked at what types of joints are used in general by the construction of the living machine. We have become acquainted with the bill of fare of possible joints. We will look in Lecture 3 at how they are actually positioned and how they work; and now we will introduce ourselves briefly to the structure and the mechanical properties of *bones*, which are connected to these joints and which present the main solid support of the body.

Bone has a huge, multifaceted mechanical strength. Its resistance to fracture differs little from the resistance of cast iron. Its resistance to crushing or breaking is higher than the resistance of an oak tree. In general, the strength of the bone is close to brass. At the same time bone structures are extremely light. The specific gravity of bone is a little bit less than two. How is this strength achieved? What substance is the bone made of?

The chemical composition of bone is not complex: it consists mainly of various calciferous salts, mainly out of phosphorous acidic lime. This substance is familiar to us, is of dead nature and is not distinguished by any special strength.

One can answer our question only by considering the internal structure of bone. The strength of bone depends not on the fact that it is made of a durable substance but on the fact that it is cleverly built. If we polish a small and very, very thin plate of bone and study it under the microscope, then we will see that bony substance consists entirely of the thinnest tubes which are threaded with very thin canals.

The aperture of the canals is so small that even a hair cannot get through them. The whole secret of the bone durability is the structure of the walls of these bony little canals.

The wall of such a little canal consists of a number of layers and every layer is a little net made of the finest fibres, which are saturated with calciferous salt. If you are familiar with reinforced concrete constructions, then you will see an amazing similarity between them and the structure of the little bone canals. Tiny fibre networks correspond to an armature of the concrete fittings and their calciferous salts correspond to cement. It is because of this combination that the strength is so high—it is approximately five times higher than the one of reinforced concrete. The fact that this is the structure of bone, made of two types of materials, can be proved without using a microscope.

If you put a bone into a white-hot oven and burn it, then all the organic fibre will burn and only the calciferous part of the bone or the 'concrete' will remain. Burnt bone like this will be very fragile and it will easily crumble into a powder. Or you can do the opposite: you can put a bone into a mild acid solution that dissolves all the calciferous salt. After such a procedure the bone will become soft like a cloth: it will be possible to wrap it around a stick.

Our organism makes not solid columns but complex trellis constructions, which resemble cranes, out of such component elements. Living bone possesses one wonderful property. It develops most of all in places and in the direction where it is affected by other forces the most, and degenerates in the places and directions where forces do not act on it. Consequently, a bone becomes a self-constructing automatic bridge. In it, all the bits gradually disappear and under-develop apart from those that are necessary for achieving the highest durability in combination with the greatest lightness.

Look at the cross section of the tibia that is depicted in Fig. 1.8. You can see that this bone has a cavity inside, a cavity that is surrounded with a thick wall. That means that the whole bone has the structure of a tube. It is built in absolutely

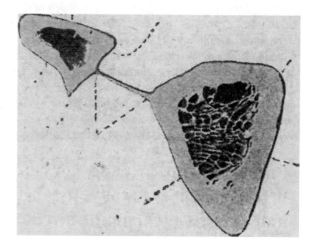

Fig. 1.8 Cross section of the bones of the lower leg. On the left the fibula and on the right the tibia (by Spalteholz)

Fig. 1.9 Positioning of
bone partitions in the head
of the femur (on the right),
in comparison with the
lines of stress in a crane
(on the left)

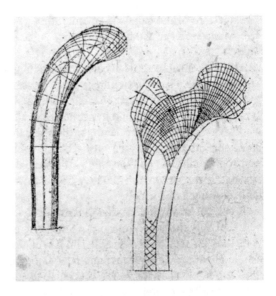

the same way as tubes of a bicycle frame and also, like the bicycle frame, it combines
strength and economy of the material. The ends of bones in which the mutual rela-
tionships of mechanical influences are more varied and complex have an even more
complex structure. They are covered over the surface by a thin continuous layer of
bony substance and inside, this bony substance creates a system of mutually inter-
secting partitions, which look like fine cells. If any of you have dealt with construction mechanics before then perhaps you have heard about so-called trajectories of
stress, which define the directions of the strongest acting force. And it turns out that
bone partitions of the ends of the bone are positioned in the direction of the stress
trajectory. Calculations have shown a high degree of correspondence here. In
Fig. 1.9, there is a longitudinal cross section of the top end of the human thigh bone;
next to it, for comparison, is given the disposition of the stress trajectories in the top
end of a crane, which is subject to forces similar to those in the thigh. The similarity
of the two drawings speaks for itself. It is interesting that in places where several
adjoining bones together are influenced by the same forces, the partitions in these
bones look like a natural continuation of each other. This is exactly what you have
in the bones of the foot. The foot is an arch that rests on the ground on three points:
the bases of the big toe, little toe and heel. The weight of the whole body presses on
them. Correspondingly, the partitions of the foot bone are positioned in the same
way as the parts of an iron arch construction would be positioned.

Lecture 2

In Lecture 2, Bernstein explains how muscles are structured and how they work.

Comrades! Last time we talked about what serves as auxiliary apparatus during movement: about bone levers and hinge joints. Both obey the movement of the motors of the human machine; they, so to speak, perform the movement but do not bring it about independently. They are the passive part. The active part of the human machine is the *muscles*.[1] The wider public know very little about how these muscles work. But it must be said that scientists do not yet fully understand how muscles work. So now we must investigate more thoroughly what muscles are, how they act and what we can expect from them. This investigation will be useful for you time and again, in order to understand how movements of the human machine take place. You have all seen muscles: if not human ones, then muscles of animals. What we know in everyday terms as meat is the muscles. If we take a lump of meat from soup, a lump that is cooked well, we see that the whole mass of muscles falls into small fibres. These little fibres represent the main component part of muscles both of humans and of animals. But this is not the smallest component part of muscles. If we look at muscles under the microscope then we see that these little fibres consist of still smaller fibrils, which are thinner than hairs, about 0.001 mm in diameter. All muscles consist of little fibres, which lie next to each other lengthwise. These fibres are gathered into small bundles, which are also very thin. If you cut a section across a muscle, you will see then simply by eye that they look like a honeycomb construction. Each thinnest fibril of the muscle represents an original *elementary motor*.

Each fibril has the ability to produce movement. This original machine, the muscle motor, however, is built in a unique way, not at all like the construction of artificial motors. Each muscle has several hundreds of thousands, maybe millions, of separate motors, which are all connected in parallel with each other and all perform work jointly. These smallest and simplest motors of the muscle substance

---

[1] Bernstein gives both the Russian word for muscles (*myshtsy*) and the homonym with its Latin root (*muskuly*).

© Springer Nature Switzerland AG 2020
N. A. Bernstein, *Biomechanics for Instructors*,
https://doi.org/10.1007/978-3-030-36163-1_3

can be examined only under the microscope. You see that this thin little fibril is covered with transverse stripes and looks like striped bathing costume material.[2] A very complex experiment with muscles, which I am not going to describe, showed that only one type of these stripes possesses the ability to make movement and the other, intermediate stripes serve as the ligament for the contracting elements. So in this fibril, only half consists of motors and the other half holds the motors together.[3] Fibrils in each bundle are fitted so neatly that these stripes are aligned with the row of fibres lying next to them. Under the microscope, it is a continuous stripy bundle, which is why the skeletal muscles are called striated. But if we look at this bundle not from the side, as I have drawn it here, but from the end or in perspective, then these stripes will look like discs. Imagine that you placed copper kopeks in columns and you alternated them with silver 20 kopek coins[4]: that is how the structure of the muscle bundle looks.

How does this construction work? In order to understand how human muscles work first of all, let's look at this in its 'cold state', not working, exactly what you would do if you wanted to study a motor: you would look at it first while it was stationary. In a word, you would get to know its construction while it wasn't work-ing, and then you would start it up and see what happens. So let us see what muscles are in themselves and how they work. Figure 2.1 depicts two bones between which a muscle is stretched. This, in essence, is the form that is repeated in all the arrange-ments of muscles of the human machine. But we cannot cut out muscle from human beings for experiments. For this, it is best to take an animal, for example, a frog. If you cut a frog muscle out and try to stretch it, then you will see that it is elastic; it resists, that is, it develops some force. The muscle stretches like a spring. If you take a frog's muscle and hang a small-scale weight from it, then it will stretch and become longer, but as soon as you take the weight off, the muscle will regain its former length. There is a difference from an ordinary spring and we will come to that later. In order to imagine the muscle spring, we will think about a simple spring to begin with. If you hang a load at the end of the spring, it will stretch the spring. What does this mean?

You know that elastic bodies, for example, a spring, have the property of return-ing to the same size as soon as you stop stretching them. In order to somehow change their form, in order to stretch a spring or elastic body, it is necessary to apply some sort of force, a scale weight in this case, and this scale weight, by the force of its weight, will stretch the spring. It is well known that if it is necessary to apply a force for some movement, this means that this movement needs work. You know, probably, that if you take a weight and let it drop to the ground, then while falling it carries out work and this work is measured by the product of the weight of the load and the height at which the load was dropped. The work is measured by

---

[2] Bathing costumes were all made of tricot, a woven fabric, at the time Bernstein was writing.

[3] Bernstein's description of the different functions of the different 'stripes' in muscles is incorrect as he was writing before the discovery in the 1950s of the importance of how myosin and actin fibres overlap, in the sliding filament theory. The theory explains why the A bands stay the same while the I bands get thinner during muscle contraction.

[4] A *kopek* is a unit of Russian currency, 1/100th of a *ruble*.

Fig. 2.1 The action of muscle on the bone lever. This muscle (biceps or the two-headed muscle of the upper arm) acts only on the forearm bones, so that squeezing the wrist into a fist, depicted here, is a whim of the artist

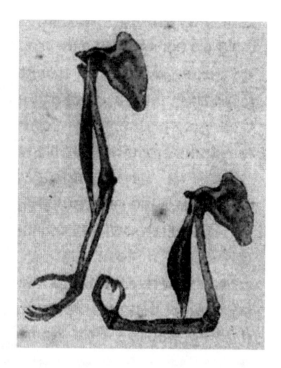

metre-kilograms. We can say that if, for example, we hang a 1 kg weight on a spring and the weight stretches it by 1 m, then it means that the weight has produced work of 1 m/kg. In order to stretch our spring to the given magnitude, we needed work to the extent of 1 kg/m. Where has this work disappeared? Where has it gone? Nothing ever vanishes in nature.

Moreover, the weight, having fallen downwards, has stopped still, and doesn't move any more. That means that work has disappeared somewhere. Where is it?

Student: This work was used in stretching the spring.

Lecturer: Of course, it has been expended in stretching the spring. And if you let it go, it will shorten again with force. That means that all the quantity of work has been turned into the hidden form of the tension of the spring. The more the spring is stretched, the bigger the store of work is hidden in it. Let's take a different kind of spring, which you know better, for example, a gramophone spring. If you stretch the spring, you will create tension in it and this work will be almost entirely returned to you by the spring when it starts to unwind. That means that the main quality of the spring is as follows: *it can be charged with work, creating tension in it, and it can return that same work back to you at any moment*, when the spring is again unwound and loses its tension. It returns the work that we imparted to it. Now it's obvious that you can stretch the spring, let's say, by hand or with a load and then you can let it shorten back. Let's see what can happen with the frog's muscle.

Here, you stretched the muscle, created some force of tension in it and charged it with work. This force of tension in it can be liquidated, if you give it the opportunity to shorten back. The muscle will contract and return part of the work. Now why part of it? This is the reason. However ideal a steel spring you have taken is, it cannot return all the work all the same, because part of the work is used in the internal friction of the particles. However good the spring is, there is always friction in it, and work is always expended on this friction. Therefore, part, maybe 95%, but not the whole 100% will be returned. The situation is worse with muscle. It is not capable of returning even 50%; it expends an enormous amount of this work on friction. And if you stretch it and give it an opportunity to shorten by itself, it will return not all the work, but only part of it. You might think that if a muscle is 50 cm long, and we stretched it further by 8 cm, then after that it would no longer be able to shorten to the former 50 cm but only to 51–52 cm. Let's take, for the last time, the same spring, where what happens is much simpler. What will happen if you hang a weight on it? Does this weight begin to stretch it? Quite right. What does the stretching mean? Let's try to analyse this, so that for the rest of your life you won't forget what it means for the weight to be stretching the spring. The weight pulls the spring downwards with a force that is equal to its weight. Let's write it in numbers: let the weight be 1 kg. How can we suspend a weight in order for it to be in balance? When we just suspend the weight on the spring, the latter cannot resist yet, because it has not been stretched yet. That means, the weight starts to fall down to where its own weight takes it and doesn't meet any resistance from the spring. As soon as it starts pulling the spring downwards, at this moment, thanks to the stretching that has started, some resistance is created which starts pulling the weight upwards. The weight pulls downwards with the force of 1000 g but the spring started to exert, for example, a force of 100 g. In the end, we get the resultant of these two forces: 1000 downwards and 100 upwards, i.e. 900 g directed downwards. In order to find out with what force the weight pulls the spring down, one has to take the difference between the two forces. You know that the more strongly we stretch the spring, the greater force of resistance it has, so that it is possible to stretch the spring to the extent that you can have any resistance you like (obviously within the limits of the elasticity of the spring!) So, we are going to stretch the spring with all effort possible, that is, not us ourselves, but the falling load will stretch it to the extent that the resistance will be equal to the same 1000 g, that is, the same as the weight of the load. That means that both forces will counterbalance. But does this mean that the load will remain at the same height? You see that the load, while going downwards with the action of the force, falls downwards faster, because the force is directed downwards all the time. The increasing resistance of the spring reduces the resultant force, but until the moment of equilibrium, it is directed downwards all the time and the scale weight does not encounter the force, which is directed upwards.

So, what follows from this? What follows, no more, no less, is that the spring and the weight, at that moment, will have a very decent speed of movement and at the point of equilibrium *they will not stop*. Moreover, the scale weight, which is not braked by anything, will go through the point of equilibrium—and that means that it will stretch the spring even more strongly. Again, the load of the scale weight all

Fig. 2.2 Stretching of a spring with a weight. The white arrows pointing upwards represent the tension of the spring; the arrows pointing down represent the force of gravity of the load; the black arrows represent the balancing force of both forces. x_1 is the point of equilibrium (the resultant force = 0)

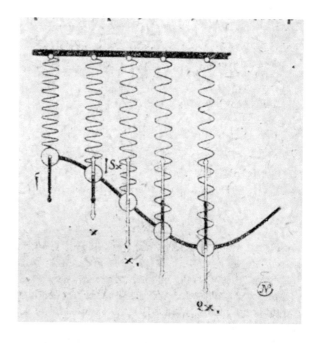

the time will be equal to 1000 g downwards but the resistance will now be more than 1000 g. Now the balancing force tries to climb upwards and this time it will be a brake for the load. The point of equilibrium is the point through which the force will go with the greatest speed; when lower than that point, it will start slowing down because the force of the spring is a brake. The weight will go the same distance downwards from the point of equilibrium as it went to above it. Here's what we get when we depict the stretching of the spring with a weight graphically (Fig. 2.2). It will go through the position of equilibrium and the weight will go to the place that is two times lower. But because there is no equilibrium at the point where the weight will finally stop, the spring will begin to start pulling the weight back. It will come back but again will pass too quickly through the position of equilibrium. In a word, if the spring is very good (no doubt you have observed this many tens of times in your life), then it will simply start to bob up and down. And finally, these oscillations will 'die out' very slowly at the level of equilibrium. The spring will stop only when all the work from the original drop of the load has been used up in friction. Until the point where it is used up, this work will be expressed in the movement of the load back and forth. And now let's return to our muscles.

The difference will be that the forces of friction in a muscle are huge. Let's now imagine what will now happen to the muscle on the sketch in Fig. 2.3: whether it will oscillate like the spring does or not. I will take several oscillations in order to see it clearly. Here is what the spring will do when we hang a weight on it and let it go. The friction inside the muscle is very great and the muscle will be stretched by our weight very slowly. It will be very difficult to make it 'swing'. Besides, it turns out that its oscillations will die out much more quickly; one or two oscillations and the

Fig. 2.3 This is how the waning oscillations of the spring look, disturbing the balance of the weight. The waning oscillation of the muscle (Fig. 2.4) looks very similar but it dies away quicker there

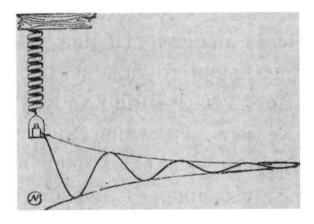

muscle has stopped. It has already swallowed up all the work of the load that made it oscillate. After a very short time, the muscle found the point of equilibrium, stopped at its new length and balanced the weight.

We now have to agree on this. When I say that a muscle counterbalanced some weight or the other, it means that the muscle tensed with a force that is equal to the force of gravity of the load that is hanging on it. To say that the muscle has counterbalanced a load of 1 kg is equivalent to saying that the muscle has tensed to the extent of 1 kg. But a muscle has one remarkable quality that none of our artificial springs have. This quality consists of a special way of charging a muscle with work. For a spring or for dead muscle we have only one method: take a muscle (cut from a frog) and stretch it. But it turns out that the muscle that sits in its place in a living organism has another way of being charged with work, a way such that it is not necessary to stretch it. This way charges a muscle using *internal processes* with no visible movements. This differentiates live muscle from those motors that we know. It is called *stimulation of a muscle*. If we send an electrical current through a muscle, even only one electrical discharge, then suddenly the muscle will pull the load upwards. You all know that the muscle shortens and there is nothing surprising in this, but from a mechanical point of view, it is an extremely amazing thing. Let a muscle counterbalance 1 kg. Now if the muscle pulls the weight upwards that means that the balance is disturbed; that is, the force that was directed upwards is outbalanced. It turns out that the muscle received tension not to the extent of 1 kg but more, let's say 1½. Where has the muscle got this ½ kg from? It is incomprehensible. Where has the muscle taken work for the additional tension from? The conclusion is that the stimulation of the muscle achieved, for example, if we send an electric charge through the muscle, charges the muscle with work, not mechanically but in some other way.

I need to bring your attention to the fact that the muscle that shortens from stimulation from an electric charge, for example, behaves in exactly the same way as the spring that had great friction. Muscle always and at all times behaves like a spring. The only difference is that this living muscle can be charged with work with the help of an electrical process.

Living muscles of the human organism are also charged with work by the means of an electrical impulse of a special kind, but this impulse is transmitted through the *nerve*. We will speak later about how this happens. In this case, a portion of work is released in the muscle, not from the mechanical stretching, but from combustion of some parts of the muscle substance. A muscle can be compared with an internal combustion engine, let's say a diesel engine; and the combustion of the fuel that releases work is taking place in it.

Let's look at this from the point of view of muscle contraction. Let's take a muscle from a frog organism again. We will cut it out together with the nerve and a little bit of the spinal cord; it won't die straight away and we can experiment with it. If we put a little tiny load on this muscle and balance it, then the gravity of the load and the tension of the muscle will be equal. If you pass an electrical current through then the muscle at first will tense more strongly, staying for just a second the same length as it was before it was stimulated. At the first moment, one or two hundredths of a second, the contraction won't be noticeable. Then in the next moment, the muscle will start contracting because its tension is stronger than the weight of the load. Note that the tension of the muscle grows together with the weight of the load. The greater the load, the greater the tension. *That means, when the muscle shortens there are two consecutive moments, two consecutive events; at the beginning the muscle tensing takes place, and then the shortening connected with the relaxing of the muscle starts.* These events are named as follows: the phase of tension (without movement) and the phase of contraction (muscle relaxation).[5]

You don't know of this little detail or it has never come into your head. We are so used to talking about muscles getting tense in movement, and it has not come into anyone's head that muscles relax and do not tense during work. When a muscle has to move, it relaxes. When a muscle is tense, it does not produce work. Why is this? It will become clear if we understand the similarity between the mechanism of the muscle and the spring. Now, let's describe the action of muscle, its contraction. We have to know the work of the motor very well in order to understand more complex things.

So how does muscle contraction take place? First of all, I have to point this out to you. We said that the muscle is stimulated and gets charged with work when it is exposed to an electric current or under the influence (also electrical) of the nerves that go from the centre—the brain—to this muscle. Science does not know exactly what stimulation of muscle is. What is important is that electrical phenomena in muscles have definitely been proved. Something else has also been proved. It turns out that if we stimulate a muscle in any way, then its electrical charge changes at this moment. If you connect the muscle to a galvanometer then you will see that as soon as the point of a muscle gets stimulated, the muscle gets a negative charge. It is important to know that here there is always some internal current, not great, only about one-fifth of a volt, and no more. If you bring about stimulation in some point of the muscle, then this will immediately disperse to both sides of the whole muscle,

---

[5] Apparently here Bernstein means isotonic contraction rather than a release of tension by 'relaxation'.

with a speed of approximately 10 m a second, the speed of a passenger train. But then it turns out that in each given point of a muscle the stimulation holds on for a period of time that is not long, for a moment, and then the muscle returns to its previous resting state. It's not possible in any way to keep a muscle stimulated for a long time. The muscle can get stimulated only for one moment; it is as if it just twitches and then calms down immediately afterwards.

If you measure what happens in the muscle during stimulation with the help of a galvanometer, and connect the galvanometer with any two points of a muscle, then this is what you get. Let's say the stimulation started at point A. At that moment, point A will be electrically negative, and point B will be positive. Some hundredths of a second later, the stimulation will manage to run from point A to point B. Then, of course, the opposite will happen: point A will become neutral again and point B, now stimulated, will be negative. The current gets the opposite direction. Therefore, the galvanometer at first goes in one direction and then in another. If we try to record what happens at a given point in the muscle, the record will show at first the resting state of the muscle and then the oscillation of the current in one direction or another and then everything calms down. This oscillation of current in a muscle that has been stimulated is called the *action current*. It always happens when the muscle gets stimulated and it is always extremely brief.

Now you ask me a question: How can it be that we can show on our muscles that they work for several minutes but you say that they get stimulated only for a moment? That's right. But it turns out that it is not as simple as this. It is clear that if a muscle gets such a short impulse of stimulation, as much as it can, in my words, receive and only that, then it cannot contract for a length of time. It can only twitch. If you conduct an electric current through the muscle or its nerve, then the muscle will act as follows. Imagine we put a scale weight on the muscle and we join to it a little writing instrument. Clearly, if this weight goes down or goes up, then the instrument will move with it. Now, if there is a cylinder next to this writing instrument covered with smoked paper, at the moment when the muscle is twitching, this instrument will record a line which will draw a curve, depicting the contraction of the muscle. In this drawing, Fig. 2.4, you see the curve of the muscle twitch. While it is stimulated it can pull the weight upwards. But as soon as the stimulation stops, the muscle which has begun to contract is in an extremely shortened state and so the weight will start pulling it down again.

How can we get the muscle to prolong contraction for a period of time? It turns out that for this, one has to stimulate a muscle not once but many times. The muscle cannot contract continuously without interruption: it can shorten only with little impulses. If they are very frequent (50 times a second), then they fuse altogether.

Now let's investigate more closely what happens with living muscle. We've established that there is no way to get a longer stimulation of a muscle than for several hundredths of a second: if we limit ourselves to one such single stimulation, then the muscle will twitch; it will twitch momentarily. You see how the curve goes, and then there is nothing. The excitation of the muscle depends on the fact that in this case, part of the substance of the muscle itself decomposes. This substance, while decomposing, frees part of its energy, which turns into the mechanical energy

Fig. 2.4 (A) The curve of the muscle action current, (Б) the curve of the twitch of the muscle, (B) the oscillation of the tuning fork which marks hundredths of a second. This is a photographic record (by Yudin)

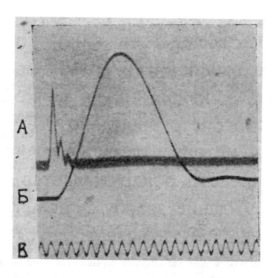

of the *tension of the muscle*. The first source of the moving force of muscle is a *chemical process*; it is a kind of burning inside the muscle. Thus, the phase of tension begins. If we manage to record precisely the contraction of the muscle on the same rotating cylinder where we are also recording simultaneously both the current that we are using to stimulate the muscle and the action current of the muscle itself, then we will see exactly how the consecutive phases of the muscle contraction follow one another. We can arrange it so that the current that we have passed through the muscle records the start of its action on the cylinder. It turns out that at the moment when the current goes through the muscle and the stimulation is taking place, the muscle is still immobile. In order to move some load from its place, it needs a force that has to last for a while. And so, for one or two hundredth parts of a second the muscle stays put and only after that it starts moving. However, it has only gone a short way when the muscle stimulation is already finishing but the muscle doesn't subside and continues moving, shortening at first and then lengthening again, which lasts for 20–30 hundredths of a second. What causes it to go upwards? After all, the stimulation has already stopped? It happens only because of the inertia, of the momentum of the scale weight that the muscle managed to receive at the first moment. It jumps upwards a little more for a short time and then comes back. If we use the same cylinder (now covered not with smoked but with photographic paper) and record the action current of the muscle as well, then things will become even clearer. You remember that the electric current, which passes back and forth through the muscle, corresponds exactly to the time of the muscle stimulation. As soon as the current has stopped, the stimulation of the muscle also stops. So if we record the oscillation of the electrical current, which started at the moment that we stimulated the muscle, then these oscillations will look approximately like this (Fig. 2.4). You see this oscillation of the current back and forth. The figure shows that the excitation of the muscle is the strongest exactly when it has not twitched yet, when it has not yet moved from its place. Why is there such a divergence in time, and why does the twitch happen after the excitation? I think that it is clear enough from what we were talking about at the beginning of the lecture.

Now imagine that at the moment when the muscle has already partly jumped upwards, we stimulate it again. This is what happens. Because the first stimulation finished very quickly, the muscle managed to jump up only a comparatively short distance, and therefore in this new position, where a second electrical charge will catch it, it will be slightly contracted. This means that in this new state, its properties will closely correspond to its properties before the first stimulation. What happens is that the muscle will jump up even higher, moreover almost to the same height, starting from the point where the second stimulation caught it. The curve will look as it does in Fig. 2.5. Now if we give a third stimulation it will jump up even further. If we give it several stimuli one after another, then it will turn out that its contraction is becoming higher, much higher than after the first stimulation. If we were to stimulate a muscle constantly, over and over again, then the curve will look like a zigzag line: the muscle will lift the weight to a height at which it will quiver for a while, and only after we stop our consecutive stimulations will it come back. If we start stimulating it about 50 times a second then it won't have enough time to go down after the first stimulus and the curve showing its contraction will look smooth; it will be giving such insignificant oscillations that it won't be possible to see them on our recording. Therefore, if you stimulate the muscle very often then it stays shortened for a long time, trembling just a little bit at the height that it achieved.

We can get a prolonged shortening of the muscle only when stimulations follow one after another frequently. If a human muscle gets 50 stimulations a second, that is, 3000 a minute, then it contracts very smoothly like this. We can do a simple experiment on ourselves. All you need to do is to clench your jaws tightly, in order to hear a low creaking sound near your ears. This sound is nothing more than the sound of a muscle that is used for clenching the jaws and is situated near the ears in the region of the temple. As soon as you clench your jaws, you will create a prolonged tension of the muscle. This muscle will alternately contract and then relax, and begins a very fine quivering, which you cannot sense by touch but which sounds like creaking. If you tense some muscle of your arm strongly, then you will notice the quivering which takes place in prolonged muscle contraction. If we put an ear to

Fig. 2.5 (I) Composition of two consecutive single contractions. (II) Composition of many contractions following on from one another. (III) Tetanic contraction (by Landois)[6]

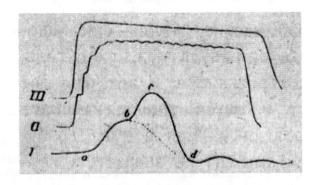

---

[6] Leonard Landois (1837–1902) was a German physiologist.

a muscle that is in a state of contraction then you will hear a low bass sound. This is called 'muscle tone'.

Physiologists have shown a very interesting thing. Human muscle never performs isolated contractions in natural conditions. In other words, the nervous system, which sends stimulation into the muscles, is able to send dozens of stimulations per second. Our nervous system cannot send one stimulation or one impulse; it can only send such rhythmical stimulations. That means that our muscles can only contract at the command of the nervous system using this complex model we have just described, which results from the fusion of very frequent consecutive stimulations. We call this *tetanic contraction* of the muscles or *tetanus*. The muscle can get such tetanic contractions only through the nervous system.

Of course you can get human muscle to carry out a single twitch artificially. If we conduct a single electric shock through any of our muscles from an induction coil, then the muscle will carry out a real single twitch. But voluntarily you can only produce tetanic contraction.

That is enough talk for now about the action of muscle that is artificially separated out. We need to go on to look at how muscles work in the conditions of the human body and find out how they are constructed and how they act. We have already spoken about the construction of muscle fibrils. Now we have to say that the fine muscle fibrils positioned neatly parallel to each other are immersed in quite a viscous liquid, which itself does not take an active part in the movement of the muscle but serves as a casing for the muscle fibres. You can imagine the general construction of a muscle in the following way, such that it is a packet with thin fibres that are surrounded with internal muscle liquid, of course in a very thin layer. Every bundle of fibres is covered in a thin casing that separates it from the neighbouring bundle. A multitude of such bundles are situated next to each other with a small space between them. The whole muscle is covered with the casing from similarly springy elastic fabric but stronger. This casing also protects it and wraps round it on all sides. At the ends of the muscle, if you take it as a whole, the transversely striated muscle fibres disappear and gradually turn into tendinous fibres. Tendinous fibres are distinguished by their extreme strength. They are passive themselves and don't take part in bringing movement about, but they serve to attach the muscle to the bones or organs which the muscle is to move. Tendinous fibres are thinner than the muscle ones; therefore, the tendon itself is usually thinner than its muscle. The word '*myshtsa*' comes from the word mouse ('*mysh*'), and this is because it reminds us of a little mouse with its little belly and thin tail. We will consider the way the muscle attaches to the organs that it has to move in one of the future lectures; for now let's just look at the drawings and consider any questions you need to ask.

Lecture 3

In Lecture 3, Bernstein discusses the movement of the upper and lower limb, taking into account evolutionary factors.

Comrades! Today, our task is to examine the separate segments of the human body and their mobility. We will be referring all the time to what we called in Lecture 1 an abridged plan but, in many cases, we will have to deviate from such a simplified system and consider the mobility of joints in all their complexity. That will be necessary especially when we look at the mobility of the arm. As you will see later on, this mobility is very varied and complicated; and because the arm has a huge practical role in any work, it is clear that we will have to study its construction with special care.

You probably remember that the upper extremity is joined to the body by means of two bones—the shoulder blade and the collarbone—and a great number of muscles. This whole combination of transmission bones and muscles is called the shoulder girdle. In the following lectures, when we will talk about muscle equipment, we will consider why the shoulder girdle has come to be constructed as it has; but at the moment we will accept it the way it is, and have a look at the conditions of its mobility.

The shoulder girdle has almost no bony fixings. One of its bones, the shoulder blade, is not attached to the torso in any way at all. It is linked to the collarbone only with a small, mobile and fragile joint. The collarbone, in its turn, is connected with the sternum with a similarly unstable joint and the latter is fixed to the spine indirectly as well, by means of the ribs. Thus, the upper arm, which has a joint only with the shoulder blade, is very distantly related to the supporting rod of the torso. Such a distant link and the wealth of intermediate joints give it exceptional mobility. As regards the stability of how the arm hangs from the body, this is achieved only with the help of muscles.

Before we look at the mobility of the shoulder, let's consider the movement of the bone on which it hangs, i.e. the shoulder blade. The shoulder blade is tightly pressed against the back wall of the ribcage by muscles and can move only by pressing close against it. That means that its mobility with regard to the rib cage is the same as the mobility of a flat plate on the table, that is, it has three degrees.

© Springer Nature Switzerland AG 2020
N. A. Bernstein, *Biomechanics for Instructors*,
https://doi.org/10.1007/978-3-030-36163-1_4

The shoulder blade can move up and down, towards the middle and to the side, and also it can rotate around itself.

Let's call the first type of movements of the shoulder blade elevation and depression, the second adduction and abduction and the third one rotation of the shoulder blade. We can easily observe all three types of movement in a living person. It is true that people rarely are able to make all these movements with the shoulder blade at will, especially rotation. Because one of the corners of the shoulder blade (Fig. 3.1) is connected with the collarbone and the other end of the collarbone is fixed to the ribcage with almost no mobility, then in any movement of the shoulder blade, the collarbone will turn in various directions with its outward end. You can observe the movement of the collarbone, if you carry out movements of the region of the shoulder joint, that is, what is referred to in everyday speech as the shoulder.

The shoulder blade has quite extensive mobility. Its lower end can move by 16 cm or more to the side and by 10 cm up and down.

Now let's get to know the most mobile of all the joints of the human body, the one that joins the shoulder blade with the upper arm. This is usually called the shoulder joint, for short. As we discussed before, this joint is attributed to the three-degree ones. In Fig. 3.2 you can see that it is a type of ball-and-socket joint. The humerus has at the top end a head in the form of a hemisphere, and the shoulder blade has in its upper external corner a hollow of an appropriate form. You can see in Fig. 3.2 that the hollow is much smaller than the head of the humerus. This means that the upper arm can swing with great range but the hollow of the shoulder blade will always have a place where it comes into contact with the large ball-like head. It

Fig. 3.1 The right shoulder blade from behind and the types of its mobility. *1*—adduction, *2*—abduction, *3*—elevation, *4*—depression, *5*—rotation inwards, *6*—rotation outwards

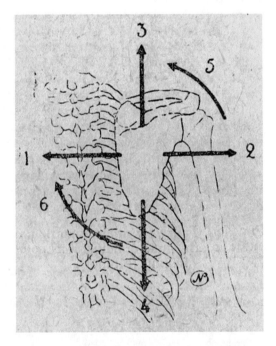

Fig. 3.2 An X-ray of the
shoulder joint. On the right
is seen the humerus with
its ball-like head and on
the left we see the scapula
with its hollow. At the top
we can also see the
shoulder blade's process
for the collar bone (by
R. Fikk)

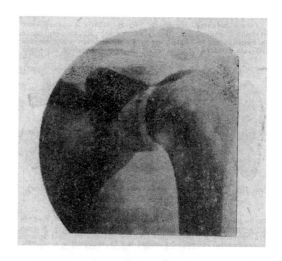

is true that there's no talk of flanges or hard braces. Moreover it is obvious that the joint capsule has to be very flexible and wide in order to make way for the great range of the humerus. That means that the securing of the joint is weak and you can sprain it more easily than any other joint.

Now let's consider the variety of shoulder movements in this joint. It's important to notice, first of all, that both shoulder blades are positioned to some extent diagonally in relation to each other, so that the plane of each of them is oriented forward and inwards. The articular surface of the shoulder blade is fixed at a right angle in relation to that surface, that is, it is looking forward and outwards. If we start turning the upper arm in this last named direction, that is, we start lifting it forward and outward, then the movement that results will be *extension* of the shoulder. The movement opposite to this, that is, lowering the upper arm in the same plane, is called *flexion* of the shoulder. [1]

I'd like one of the comrades to come out here to the lectern and to show how he produces extension and flexion of the shoulder just in the shoulder joint alone, i.e. the shoulder blade should remain immobile. I will also ask two other comrades to observe the first person's back and to check whether the shoulder blade is moving or not.

Students: He is moving his shoulder blades.
Lecturer: You are not doing what I asked you to do. Extend the shoulder while the
                shoulder blade is immobile. Notice that our volunteer cannot do that.
                However, it doesn't mean that there is something wrong with his mobil-
                ity; perhaps none of us can do that. However, this movement is quite
                possible. With our example, you see how crudely and imperfectly the
                human being controls the movements of the most important parts of his

[1] It appears that here and in the discussion later in this chapter what Bernstein refers to as 'shoulder extension' and 'shoulder flexion' now refer to the opposite movement.

body. Now try to keep the shoulder blade of our volunteer immobile, holding it still by its lower corner. Determine now to what extent he can extend the shoulder without moving the shoulder blade. Here he has taken his arm to the horizontal position. Look, when he extends the arm even higher, the shoulder blade starts to turn and you cannot hold on to it any longer. Here, it is not the lack of knowledge of how to use the muscles that is to blame: he has simply reached the limits of the mobility of his shoulder joint. If you try to produce a movement of the shoulder in all other possible directions, then it will turn out that the limits of the mobility of the shoulder in each direction are only 100–110 degrees; anything further and the shoulder blade starts moving. If you fixed the shoulder blade completely so it is immobile and then made the shoulder take all the extreme borderline positions one after another then the end of the upper arm would draw a figure that resembles a circle (Fig. 3.3). When the shoulder blade is immobile, the shoulder can move its lower end to any point inside this circle, but it is not in a position to go beyond its borders. It looks like this circle is connected somehow, in a way that cannot be changed, with the shoulder blade, because the borders of mobility of the shoulder are always defined in relation to it. That means that in order to put the shoulder in some position other than inside of this magic circle one has to move this circle in space. And this can be done only when you turn your shoulder blade in a corresponding way (Fig. 3.3). Now we understand how the movement of the shoulder blade widens the limits of the mobility of the shoulder. Which movements of

Fig. 3.3 How the turns of the shoulder blade (A) influence the borders of mobility of the shoulder (Б) (by Mollier)

the shoulder blade widen the borders of extension and flexion of the shoulder? We won't guess; we will be better to check on a live example. Make our volunteer flex the shoulder and explain what happens to the shoulder blade.

Students: The shoulder blade is turning.

Lecturer: When extending the shoulder, the shoulder blade makes a turn inwards, and when flexing, outwards. This is understandable; after all extension and flexion are made around a horizontal axis that is directed forwards and inwards; that means that the auxiliary movements of the shoulder blade are those which are made around an axis which has the exact same direction. You can see that with the help of the shoulder blade, the shoulder can be extended by approximately 60 degrees above the horizontal line.

Let's have a look at another type of movement of the shoulder. When we were extending the shoulder, the end of the upper arm was moving in the vertical circles. Now let us consider its movements along the horizontal circles, that is, turns of the shoulder along a vertical axis. This movement is called *adduction and abduction* of the shoulder. It doesn't matter at which height it is made. Determine now which movements of the shoulder blade can help this type of displacement of the shoulder. You can see that adduction of the shoulder is accompanied by the abduction of the shoulder blade and abduction of the shoulder by adduction of the shoulder blade. Here, the participation of the shoulder blade widens the mobility of the shoulder by about 30 degrees.

We have considered and named the movement of the shoulder around two axes that are perpendicular to each other. Evidently, there is another axis, a third one, which is perpendicular to the first ones. Movements around this axis will be directed forwards and inwards and backwards and outwards: this is exactly the movement of the shoulders that is made by coachmen when they are cold. This third type of movement is called *anteroversion* (forward and inwards) and *retroversion* (backward and outwards). Can you guess what the shoulder blade will do during this movement?

Students: It will turn around the same axis.

Lecturer: Yes, that's true but how will it happen? Let's invite one of our skinny and weak comrades to come out. I make him do retroversion of the shoulder by force. What is happening with the shoulder blade?

Students: The lower corner of it is bending back.

Lecturer: Yes, and look how far. I can easily put three of my fingers under the shoulder blade. In muscular people, the muscles get in the way of it bending back in this way and so the extent of retroversion is somewhat less for them.

Thus, we have learned how to describe most of the movements of the shoulder. Let's experiment on ourselves using some examples. When a person does overarm swimming what movement does he/she make with his/her shoulder?

Students: Adduction and abduction.

Lecturer: Not exactly. The movement is close to adduction in the swing up but
when you use your arm as an oar, then it is almost pure retroversion.
The second good example of retroversion is the movement of the right
shoulder in the swing stroke with a sledgehammer. What are the movements of the shoulders during walking?

Students: Flexion?

Lecturer: No. Because shoulders during walking move straight forwards and
backwards: that means that this movement will be something intermediate between flexion and retroversion.

So, we have studied movements of the shoulder in relation to three mutually
perpendicular axes. It would be possible to account for all the possible movements
of the shoulder in its three-axis joint with these three types of movements as intermediate forms. To explain how to do it would take too much time; better if we can
single out some of the intermediate forms in a special group and give it a special
name. I'm talking about the movement of the shoulder around its long axis, which
we call rotation of the shoulder inward and outward.

The general limits of mobility that the shoulder gets, thanks to the mobility of the
shoulder blade, are great. Each of you can check this on yourself; these borders are
depicted very graphically in Fig. 3.4.

Now, let's consider the lower joints of the arm. In the first lecture we mentioned
briefly the hinge joint between the upper arm and the ulna. We need to add very little

Fig. 3.4 General limits of
mobility of the shoulder
(by Mollier)

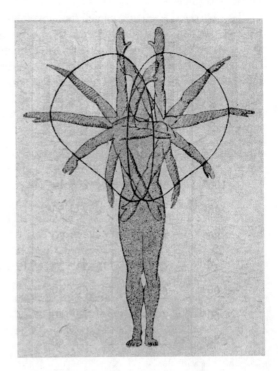

to describe this one-degree joint. The limits of its mobility are about 140 degrees and because the strong muscles and ligaments don't allow big movements, the mobility of the elbow in weak people, also in children and women, is greater than in strong men. There are women and children who can extend the elbow further than a straight line with the shoulder (hyperextension of the elbow).

The way the upper arm bone joins another bone of the forearm—the radius—is much more curious. In Fig. 2.4, one can see that the humerus has at the lower end a small ball next to the pulley. The top end of the radius ends with an exactly matching spherical hollow. One would have thought that in this kind of ball-and-socket joint structure one could expect a mobility of three degrees between the radius and the upper arm bone. Reality disappoints our expectations.

The radius ends below by the base of the thumb; in this particular place one can feel its protruding end. Now follow the movements of the radius when the shoulder is fixed and, orientating yourself by this protrusion at the lower end, define which kinds of movement the radius can perform and how many degrees of mobility it has.

First of all the radius, together with the elbow bone, can take part in flexing and extending the elbow. This is the first type of its movements. Secondly, it is also mobile in relation to the ulna and lies alongside it. Turn the wrist with your palm upwards and again downwards; you will see that in this movement, the radius crawls around its neighbour, the ulna. If we follow it exactly, it will turn out that the axis of this movement lies along the forearm almost longitudinally: it is directed from the ball at the lower end of the humerus towards the inferior end of the ulna. This axis, when the elbow is straight, is an exact continuation of the axis of rotation of the shoulder.

The second type of the movement of the radius is called *pronation* and *supination*. We make the movement of supination when we are fixing a screw into the wall, and the opposite movement, pronation, is loosening the screw.

We have two types of movement now. Is there a third one? If there is, it should be bending the radius to the side so that its lower end moves away from the ulna. But it is attached in the wrist at this lower end; therefore, the radius does not use its third possible movement; in fact, it has a mobility of two degrees. The movement of pronation and supination when the arm is straight is always accompanied by rotation of the shoulder, because both movements share a common axis. In this case, the same thing happens as in the case of the shoulder and shoulder blade; again, a human being cannot separate the two similar movements. However, when the elbow is half bent it is easier to distinguish both movements; indeed, pronation in these conditions will be expressed as before by rotations of the forearm around the longitudinal axis and rotation of the shoulder will make the forearm turn around it like a spoke in a wheel. Perhaps the movements of the forearm are depicted more clearly on the model that is drawn in Fig. 3.5. From this drawing it follows that the radius, strictly speaking, doesn't have to be joined to the humerus; it would have had the same form of mobility if it was joined with just the ulna, with a one-degree joint. The radius in fact is simply an offshoot of the wrist which sprouted backwards long ago, and in many mammals it in fact doesn't go as far as the humerus.

The hand is linked only with the radius. At the base of the hand there are two rows of small bones of irregular form, the so-called wrist. Between the upper row and the radius and also between the two rows, there are two joints that are posi-

Fig. 3.5 Model which
shows the way the radius
(Л) is fixed to the ulna (л)
and the humerus (П) (by
Braus)

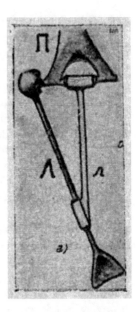

tioned one after the other: the radiocarpal joint and the intercarpal joint. The first one is ovoid (two degrees of mobility); the form of the second one is quite incoherent and it is difficult to foresee anything about its mobility just by looking at it. Experiment shows that it is also a two-degree one; therefore, we will be considering the mobility of both joints together. If I hold the forearm of the volunteer by its lower end then you will see that the hand can only rock in all directions and it cannot turn around its longitudinal axis. Check on yourself in your free time what the boundaries of mobility of your hand are in the radiocarpal joint.

The hand itself together with the fingers has many small joints (the hand consists of 27 small bones) and we won't have to look at the movement of all of them; this has already been done in detail in my book, *General Biomechanics*. Here, we will single out briefly what is most important for us.

The joints between the hand and fingers (metacarpophalangeal joints), which protrude as the knuckles on the back of the hand are, again, all ball joints. The metacarpal bones, which lie in the fleshy part of the palm and are almost immobile in relation to each other, have at their far end ball-shaped heads. The main phalanges of the fingers have sockets, which correspond. Moreover, the mobility of this joint is only two degrees (the same as between the radius and the humerus). You can actively, with the help of the muscles, flex and extend, and adduct and abduct, each finger: it is true that the two latter movements are within very narrow limits. We can't rotate a finger around the longitudinal axis; we do not have appropriate skills or appropriate muscles for that.

The metacarpophalangeal joint of the thumb is also a ball-and-socket joint but it has very little mobility. The great mobility of the thumb depends on the mobility of its joint with the hand. You will remember that its joint with the wrist has a saddle form (Fig. 1.6). The mobility of the carpometacarpal joint of the thumb is versatile;

it is thanks to it that the thumb can be opposed to the other fingers. This small fact of the slightly greater mobility of one of the bones of the hand has had a decisive meaning for the fate of all of humanity. One can say with assurance that it is thanks to this structure of the thumb, which can bring about a great variety of grasping movements, that human beings first learned to use implements and instruments. Apart from upright gait, perhaps, no other biomechanical fact had such a decisive cultural-historical meaning for human beings.

On this, we will finish with the hand joints. We will be talking more about the movements that are possible for the hand later, in connection with discussion on the muscles of the hand. Now let's talk about the mobility of the head, neck and torso.

Perhaps it is worth saying a few words about the mobility of the lower jaw. It is connected with the cranium by two whole joints, which are situated on each side of the cranium under the zygomatic arches. It's difficult to judge by the form of these joints what the mobility of the lower jaw is like. We need to do a test. Describe yourselves, how your lower jaw can move.

Students: One can open and close the mouth.
Lecturer: That is one degree. Is there anything else?
Students: Forward and backwards.
Lecturer: Yes. That's one more degree. Is that all or is there something more?
Students: That's all.
Lecturer: What about movements to the side? As you can see the two joints of the lower jaw give real three-degree mobility. You can put the middle fingers of both hands on your cheeks just under the front ends of your ear lobes; in this place you will feel the protuberances of the joints of the lower jaw. Feel for yourself what happens to them during movements of the jaw.

The head represents an interesting case of mobility. Its mobility is huge but depends not just on one joint but on several, which lie one under another in a chain-like shape. The first of them, which links the cranium with the first vertebra of the neck, belongs to the ovoid type; therefore it gives the head two degrees of mobility in relation to the neck. As its cavity is situated horizontally, rocking in it about any horizontal axis, that is, to the right to the left, forward and backward, is available to the head. Turns of the head to the sides are performed in a different way. The first neck vertebra is in the form of a ring, which is situated on the next lower vertebra, which is equipped with a spike that sticks up. The first vertebra is situated on the second as on an axial bearing, and can, together with the cranium that is supported by it, turn around the spike in that direction of the third rotation, which the cranium lacks. This means that the mobility of the cranium in relation to the axial bearing is three degrees. The mobility of the vertebrae between themselves depends on the pliancy of the intervertebral cartilage cushions, which we mentioned before. But apart from these cushions between the vertebrae, (more accurately between the vertebral arches) there are real joints here, which increase the firmness of the links between the vertebrae but on the other hand reduce their mobility. In general, three types of movement in the spine can easily be distinguished: bending forward and

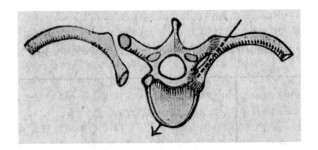

Fig. 3.6 The joint between a vertebra and the ribs. The arrow shows the axis, around which the rib can rock (by Mollier)

backward, bending to the side and twisting. Normally, the spine has several curves, which are preserved in the quiet standing position. In Fig. 1.1 it is clear that the cervical and lumbar parts of the spine have their convexity to the front and the thoracic concavity to the back. The property of mobility of the spine is such that it easily increases curves that exist in it, and very unwillingly straightens them out. Therefore, when we incline the spine backwards we bend the shape of the neck and the lumbar part but the thoracic part hardly changes its form; when we flex it forward, the opposite happens; the thoracic part bends whereas the neck and lumbar parts hardly straighten at all. The mobility of the spinal column when bending to the side and in twisting is most strongly expressed in the neck part and most weakly in its lumbar area.

Twelve pairs of ribs are joined to the spine in its chest area. Figure 3.6 gives you an idea about the construction of a joint between the spine and a rib. As you can see, there is a dual joint for each rib. Therefore, whatever the form of joint surfaces is, the rib can only swing around the axis that goes through the centres of both joints. Each rib is connected to the spine with one degree of mobility. The axes of the joints are situated in such a way that the front ends of the upper ribs can move up and down and the front ends of the lower ribs, as well as that, can also move to the side. That is what happens in inhalation and obviously it facilitates the expansion of the ribcage. The front ends of all ribs apart from the two lower pairs are connected between themselves with cartilages with the help of a bony joist—the sternum. It is apparent that in the movements of breathing, these rib cartilages yield to pressure in different ways.

Now, let's have a look at the leg joints, but we can't do this without some preamble. The fact is that bones, joints and muscles perform a role not just in movement but primarily in support. This circumstance has primary importance specifically in the way that the girdle of the lower extremities (otherwise called the pelvic girdle) is connected with the torso. Let's have a look at our four-legged ancestors again.

In four-legged mammals, the job of all four limbs is to serve as supports to the spinal column, which is arched between the fore and hind limbs like a bridge. In the original plan, both pairs of limbs are fixed to the props of the bridge in the same way: they are constructed, both here and there, like trestles, and the spinal column is wedged between the divergent legs of these. In the pelvic girdle, such a construction is revealed very clearly: two surfaces by means of which two hip bones are joined with the spinal column and face upwards and towards each other, and they squeeze the vertebrae between each other, like the keystone of an arch. The pelvis of a four-legged creature is indeed an arch.

Just have a look for a moment at the mechanics of the construction of arches. In an arch, gravity presses from above on its middle part (the arch keystone), and the pressure is transferred further on both sides to the supports of the arch. If we imagine that in place of the keystone there is a hinge and both halves of the arch are solid, then the pressure on the hinge from above will be transferred to the legs of the arch in the form of a separating force. The legs will want to fall apart to the sides. In order to stop this, the legs of the arches in building constructions are dug into the ground firmly. It figures that in an animal even when it stands as if 'rooted to the spot', the legs aren't actually dug in anywhere and have to be strengthened in a different way. This strengthening is achieved by the fact that between the legs of the arch of the hip, just as under the keystone there is a firm tendinous tension brace.

The shoulder girdle is constructed differently. It cannot directly wedge the spinal column, because the framework of the ribcage, squeezed in between the four limbs, prevents it from doing this. If we imagine that the fore- and hindlimbs have the same length, it will turn out that the spine—the place of strengthening the hind limbs—is higher than the tops of the limbs; it is quite the opposite with the sternum, to which the forelimbs ultimately have to be fixed, as it lies lower than their top ends. Therefore, at the same time as the hindlimbs support the spine like an arched bridge, the forelimbs have to support the rib cage as a kind of hanging construction. That is why between the forelimbs and the torso there is no direct solid connection. Their connection, like in any hanging construction, is exclusively soft and is made by muscles and tendons. We will talk about it in the following lectures, when we come on to the muscles.

You see how logically the pelvis of the four-legged mammal was adapted for the role of support. As far as a human being is concerned with his/her upright gait, all this structural consideration has gone down the drain, to a large extent. It is true that a vault or arch performs its task only under the condition where the load of the arch, its fulcra and the lower tension braces lie one under the others. Now imagine that a four-legged creature has stood on two hind legs and turns its body, including the pelvis, by 90 degrees, so that the arch turns out to be lying flat. It's clear that in this position, it cannot do what it was intended to do: it has had to turn back.

And here a series of indirect, circuitous attempts to mend the broken logic of the pelvic girdle begins. Restructuring of the pelvis in a human being in comparison with four-legged mammals gives the impression of a construction done in a hurry, without a plan and/or any calculation of what is needed for maintenance. Firstly, the spine in the lumbar area bends forwards steeply: or better to say the pelvis, together with the sacral part of the spine, steeply deviates backwards, trying to take its former position of the arch. In Fig. 1.1 from the first lecture one can see, besides this, a huge hump of spine, which bulges inwards in the pelvis. This is not to mention what difficulties this unsuccessful restructuring has brought about for the mechanism of giving birth. The human baby's skull is bigger than that of four-legged creatures but the passage for it has turned out to be curved and tight.

Even this turning of the pelvis does not ultimately improve the situation. The tension brace of the pelvis has not managed to get under the place where there is the most pressure. Then, because of this tension brace, which turns out to be in front, the tension brace at the back between the pelvis and the sacrum has developed more strongly. These tension braces have killed off the mobility of the sacrum and have brought in a new hindrance for the act of giving birth. As a result of all these unsuccessful changes, the human pelvis is quite a complicated construction and we are not going to touch further on its static properties. It was important for me to give you a notion of mechanics of the development of the pelvic girdle, as it is in humans.

On both sides of the human pelvis there is a joint cavity for the femur. I have already mentioned that the hip joint is a ball-and-socket joint (Fig. 1.5) and therefore as regards the question of its mobility we would need to repeat most of what we have already said about the shoulder joint. The difference between the two is mainly due to the fact that the mobility of the hip joint is significantly less. There are two reasons for this.

Firstly, the shoulder blade area of the shoulder joint is very small, and therefore the head of the humerus can slide along it within wide limits without meeting flanges. The hip socket of the hip joint seizes more than the whole of the hemisphere. Here is the only time in the anatomy of human joints where the head of a joint bone (the femur) is really captured tightly. The wide head of the femur is covered with cartilage, approximately three-quarters of the whole ball; that means that only a quarter remains free for its mobility, that is, 45 degrees to each side or as a whole, 90 degrees in each direction.

The second reason is that the whole shoulder girdle helps movements of the shoulders with its own movements: both the shoulder blade and clavicle. The human pelvis is completely deprived of any internal mobility; the femur can use only the borders of mobility, which the hip joint gives it.

The field of mobility of a femur is situated in such a way that it can move forward and out to a decent extent and very little backward and inward. Due to the fact that the borders of mobility are defined to a large extent by the pliability of the joint capsule, which stretches in extreme positions, it turns out that when standing and when the femur is near the border of its mobility at the back, the joint capsule of the hip joint is significantly stretched. It being stretched in this way turns out to be a useful thing. The centres of hip joints would be somewhat in front of the place where the torso is supported on the hips. Therefore the torso, because of the force of gravity, tries to fall backwards. This toppling over is prevented by the tension of the ligament. Figure 3.7 shows how this ligament forms a spiral around the top of the femur and in this way helps balance the torso. We won't talk much about the knee joint. We have already described its original structure with its concave cartilage bands. All there is to add is that human leg joints have become especially massive since the time when two legs started taking the entire load that was formerly shared by four legs. Because of this the knee joint is very wide—it is the most bulky of all human joints—and can withstand big loads. Despite the springy cushions, it has almost exactly one degree of mobility: flexion and extension. But the ligaments of the knee joint which are intended to be very strong when standing, that is, when the knee is straight, relax

Fig. 3.7 The right hip joint from behind, with all its tight-fitting muscles removed so that one can see the ligament, which forms a spiral (by Spalteholz)

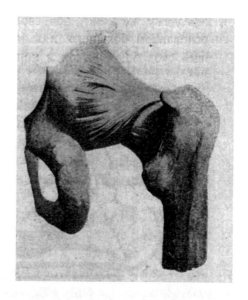

Fig. 3.8 The arch made by the bones of the foot and its muscle and its tendinous-muscular tension brace (by Mollier)

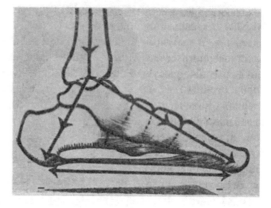

somewhat when the knee is bent, and then the joint acquires one more degree of mobility—the rotation of the lower leg around its long axis. In the lower limb, there is no crossing of the two bones of the furthest section (and why we don't have this was discussed in the first lecture) or anything similar to the mechanism of pronation or supination. In humans, the second bone of the lower leg—the fibula—has in general little mechanical significance. It is very thin; it cannot hold the weight of the torso and if the bigger bone snaps usually the small one breaks with it straightaway. This small fibula does not have any link with the knee joint and is linked to the ankle joint only indirectly. It is clearly a supernumerary member of staff.

Before we talk about the joint between the ankle and the foot we have to say a few words about the mechanical structure of the foot itself. The foot, like many of the mechanisms already considered, represents an arch. It will be easiest to see its structure in Fig. 3.8. In fact it is not one but two adjacent arches. The top for both arches is the same bone, which is called the talus. The support at the back for both arches is also one thing: it is the calcaneus, mentioned already, which ends behind with a big bump protruding far backwards. But both arches divide towards the front end: one ends with the base of the big toe and another one with the base of the little toe. The first arch is higher and springier; it leaves a print (if you step with a wet foot on the floor) only with its ends. The outer arch is flatter and always leaves a full mark. The springiness of the foot in walking and standing depends, mainly, on the arch of the 'big toe'; it defines what we call the instep of the foot.

A strong plate of ligamentous fibre, which lies on the side of the sole of the foot and which links the calcanean bone with the bases of the toes, serves as a tension brace of the arch of the foot. The strong muscles that are situated next to it help this as well.

The whole foot, like the hand, consists of a lot of fine bones, but we can confidently leave on one side the mobility many of them have relative to each other. We are interested only in movements in joints between the talus and the bones that are next to it.

At the top the talus is linked with the tibia (the ankle joint). It is an actual hinge joint with one degree of mobility; with its help, the foot can turn around the horizontal axis, that is, move the toes straight up and straight down. We call this movement flexion and extension of the foot.

Below the talus is linked with two bones of the internal arch of the foot. The weight of the body presses on the outer arch not directly but via the internal arch. This lower joint (incorrectly called the lower ankle joint) has also one degree of mobility. Its axis goes diagonally; with its help the foot turns inwards and outwards. When you are standing in the usual position for working with a chisel or file, the (front) left foot moves in the top ankle joint, that is, it flexes and extends; the back (right) leg makes movements in the lower joint. The latter movement is often called pronation and supination of the foot. Altogether, therefore, the foot has two degrees of mobility in regard to the lower leg.

In conclusion, let's have a quick look at what the joints that we have investigated in this lecture look like on an abridged plan of the body (Fig. 3.9). The first joints of all four limbs (shoulder and hip) are three degree joints—these are indicated with double circles in Fig. 3.9. The second joints, elbow and knee have one degree. They can only flex and extend. The third type of joint, wrists and ankle have two degrees. Finally, somewhere in the expanse of the second section of the limbs (the forearm and lower leg), one more degree of mobility is concentrated: pronation and supination of the hand and longitudinal rotation of the lower leg. This last movement is easier to refer to the third type of joint. Then the joints in our scheme will be positioned very symmetrically and correctly: both in legs and arms the first and third types of joints will have three degrees of mobility each, and the middle joints—one degree. Finally the head, as we remember, also has three degrees of mobility in relation to the neck, which consists of two degrees of the top joint and one of the lower joint. When studying human movements we will constantly take into account this slightly simplified distribution of mobility.

Fig. 3.9 Abridged plan of
the body. See explanations
in the text

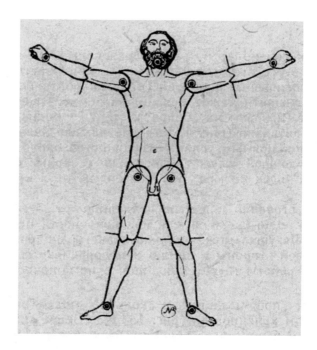

In the next lecture we will turn to a consideration of how the hinged mannequin
of the human skeleton that we have now described is equipped with muscles and
how it uses them.

Lecture 4

In Lecture 4, Bernstein explains how muscles work as levers and tension braces in joints with different degrees of freedom.

Comrades! Probably you have all seen anatomy pictures depicting a person without skin, covered with various muscles going in all directions. But maybe you never pondered why the muscles are distributed in this way and not another way and why their outlines are so strange and confusing. But, this is the question that arises inevitably for any student who, when starting the anatomy of muscles, sees before them from the first steps the necessity for hard cramming. It is natural to ask whether there isn't some kind of logical law in all this confusion of varied muscular forms, which will give an opportunity to win through here not only by using memory but also by understanding? Today we are going to try to deduce such laws and to reveal the biomechanical meaning of the distribution of muscles in human body.

In the second lecture we talked mainly about the qualities and methods of work of the simplest muscle motor—the muscle fibril. You will remember that each of these fibrils, which has a smaller diameter than a thin hair, is a real complete motor which is supplied with its own nerve and is capable of contracting quite independently. But human muscles do not just represent a pile of fibrils like this, put together without any order and sense. It is the opposite: groups of fibres are collected in larger organised units—that is what we usually call muscles—and which are organised in every case in a very different way. Let's first consider the main types of arrangement of muscle fibres, which we will meet here.

Even comparatively small groups of muscle fibres that are placed next to each other are combined together with little elastic cases, forming in this way the primary muscular units. Usually several of these units are collected together again with a shared covering; finally, each whole muscle is covered on the surface with a similar elastic cover in the form of a sleeve; in some cases this sleeve is strengthened further with particularly firm tendon fibres. The ways and the orders in which the primary muscle units are gathered into a whole muscle are quite varied.

© Springer Nature Switzerland AG 2020
N. A. Bernstein, *Biomechanics for Instructors*,
https://doi.org/10.1007/978-3-030-36163-1_5

Imagine one muscle fibril of 10 cm in length. We will imagine that such a fibril could lift the weight of one grain of sand to a height of 5 cm. As work is measured by the product of the weight of the load and the height to which this load has been lifted, we have defined with the given data something that can be called the capacity for work of the muscle fibre (I don't give the exact figures, and what is more, very variable figures, because we won't need this in future). Imagine now a second fibril of the same dimensions, structured in the same way as the first. One should think that its capacity for work will be expressed by lifting a grain of sand like this to a similar height. It is clear that if we put the two fibrils together, then both together they can lift a load equal to the weight of two grains of sand to the same height (5 cm). From this we can conclude that if you unite fibres in a parallel bundle, put them side by side, then the load that this bundle can lift will be increased, but the height to which it will be lifted will stay the same as for each separate fibre. One thousand fibres of the same size and similarly organised will lift a load of one thousand grains of sand to a height of 5 cm. We can express this as follows: the lifting force of the muscle bundle (the force, after all, is measured exactly by the load) is proportional to the area of cross section of the bundle.

Our original fibril carried its grain of sand to the height of 5 cm, i.e. half of its original length. Let's take another fibre with same properties, but its length is now 20 cm. Because this fibre has the same properties as the previous one, its weight-bearing capacity should be the same—one grain of sand. Because its properties are the same, it can carry its load to the height of half of its original length. This latter quantity, however, in this particular example, is 10 cm. Therefore, a fibre which is twice as long will also produce an increase in work capacity two times, but this time, not because of the increase of the load, but because of the increase in the height to which it can lift. We can generalise this second observation as follows: the height of the lift is proportional to the original length of the muscle. From this we can draw a conclusion (approximately) as follows: the product of the load and the height of the lift are proportional to the work capacity of the muscle, and the product of the area of cross section of the muscle and its length are approximately proportional to its size. That means, in turn, that the work capacity of the muscle is proportional to its size—in other words, the quantity of muscle substance that is contained in it.

The work that is required from the muscle does not always have the same character. Sometimes, it is required to lift a weight up a short way or simply to hold a significant load in balance. However, sometimes it's the opposite; it is a question of lifting a small weight to a considerable height. According to these opposing tasks, the muscles of the human body can be roughly divided into two classes.

One of these classes is that of short and fat muscles, which have great strength and small scope of action. In the simplest case, such muscles look like wide plates of short fibres positioned in parallel. These muscles are situated generally in those areas where the steady constant tension of a considerable force is required. A muscle of this type (the rhomboid) plays an important role in the suspension of the shoulder blade and shoulder girdle. Sometimes, the positioning of fibres of such muscles changes slightly: they are short as before and parallel to each other but they are positioned diagonally. They start on some long bone with one set of endings and

Fig. 4.1 Muscles of the neck on the right side. The muscle which inclines the head (sternocleidomastoid muscle) extends obliquely along the whole length of the drawing (by Spalteholtz)

the others finish fastened into a long tendon in a form of a cord, which unites in its end the strength of all separate fibres. Thus, the whole muscle looks like half a feather. However, there are also paired featherlike muscles.[1]

The other class of muscles consists of long muscles that are long and not fat. The simplest example of this kind is a muscle, which connects the mastoid process of the cranium (under the earlobe) with the sternum. This muscle is clearly seen on the side of the neck if you turn your head to the opposite side and also tilt it forward (Fig. 4.1). This muscle is 20–25 cm long and consists entirely of long parallel fibres. The way it works is completely different from muscles of the first class: it is not very strong and one can easily overcome its force if one turns the head with the hand; and moreover, one can hurt it with a careless movement and cause it to stretch (to 'wring the neck'). But then the scope of its movements is great. Thanks to this pair of muscles the head turns very easily.

In various circumstances these long and slim muscles take different forms as well. Here, it seems that it all depends on the place where the muscle originates and where it is attached, that is, on secondary circumstances. Very often, such muscles have a form of spindle or a fan, whose fibres start on the wide surface of the bone and then are gathered from different sides into one very fine final tendon. Among such muscles there are also more peculiar forms with two heads or two bellies, or with tendinous crosspieces in the middle, but we are not going to look at these details now. The general picture is that one can compare the muscles of the first class with a goods steam engine, with wide cylinders and small wheels, designed to move slowly and pull a great load. The muscles of second class are more like a type of express steam engine.

---

[1] Bernstein may not have known of the force-multiplying function of the diagonal arrangement of muscle fibres in pennate muscles such as rectus femoris in the quadriceps and the deltoid muscle.

However, this simple and elegant division is not defined so strictly in life. You will remember that the muscle's work capacity depends on its size and not its form. You know from mechanics that one type of work can be turned into another one, without any losses, and besides that, with simple mechanical means—for example, with a lever. If we exert a great force on the end of the lever that is closer to the fulcrum, then it is possible to balance it and even to overcome it with the help of a small force, if the force is exerted far enough away. With a straightforward lever, the balance, that is, the balancing of two forces, is achieved when the products of these forces and the lengths of the arms of the lever (so-called moments of force) are equal. Besides when a small force makes its arm go a long distance, then a great force that equalises it on the other end moves its arm only a small distance. So the obvious conclusion is that if we join two muscles, of the different classes, to the same lever, then one of them will be able to balance the other perfectly in an expedient situation and that means that it will be able to substitute for it. From this, in the application to muscles, we can see that there will be this consequence, which is of the utmost importance; two muscles can entirely, and in all respects, substitute for one another, if their moments in relation to the axis of the given lever are equal.

Bones serve as a lever for muscles. Therefore, one can judge what a certain muscle is capable of only when one takes into consideration its size and form as well, and the arm of the lever on which it is acting, that is, the distance of the point where the tendon is fixed from the axis of the joint. In Fig. 4.2, you see that a short and fat muscle, in a suitable position, can substitute fully for a long and thin muscle.

Imagine that a 1 kg load is placed on your hand. Of course, each of you is able, by flexing your elbow, to lift not only 1 kg but probably 10–15 kg. The length of the forearm in an average normal person is about 25 cm. The distance between the wrist

Fig. 4.2 These two (diagrammatically depicted) muscles are completely equal to each other

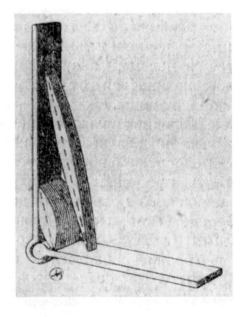

joint to the centre of the hand is about 10 cm. This means that the length of the forearm lever to the point where the load is situated is about 35 cm in an adult man. Now the second arm of the lever, which goes from the point of rotation of the lever to the place where the muscle is fixed, cannot be more than 5 cm. That means that it is seven times shorter than the arm that is holding the load. Therefore, we have here a second-class lever. Its fulcrum is in the elbow joint; one force acts at a distance of 5 cm and another at a distance of 35 cm. Let us say that the force of the load that acts downwards is 1 kg. Now the question is this: What must be the force of the muscle that would balance this 1 kg? Here's a little task for you. One arm is 5 cm, and the other is 35 cm. What force should we have here?

Student: 7 kg.

Lecturer: Imagine that one of you holds in your hand a weight of 1 *pood*.[2] A strong person can do that. Let's count it as 16 kg. This corresponds to the strength of the tension of the muscle of 112k, approximately 7 *pood*. That means the muscle as a whole is able to develop much more tension than we are used to thinking it can. Let's solve another little problem related to this. To do that we will have to simplify the proportions which exist in reality. So let's take the same elbow joint. I will allow myself to draw it here very schematically. In this joint, not one muscle but two work in co-operation. One of these muscles—the two-headed one—lies more superficially. I draw it schematically with an arrow. The other muscle, which is very difficult to feel on oneself, lies much more deeply, but is positioned, approximately, parallel to the first, in the same way as another arrow which I am drawing. I'm depicting them as a big and a small arrow. It should be mentioned that a human cannot tense one muscle separately and leave the other muscles relaxed. In future lectures, we will see how badly we use our muscles and how little dexterity we have in this sense. Also, we cannot use biceps and the internal arm muscles separately.[3] (Of course, if we applied electric conductors—electrodes— to these muscles, and pass an electric current through them, they would be able to contract separately). Let's suppose, however, that we ourselves can manage to use one or another muscle separately. Now let's solve the following problem. We'll take it that the arm of the lever, with which the internal muscles of the shoulder are connected, is three times less than the lever arm of the biceps muscle. How many times more strongly or weakly than the long one should the short muscle get tense in order to balance the same load?

Student: Ten times more.

Lecturer: You think ten times? Let's have a look at this. The lever arm of the load is still 35 cm. The arm of biceps is 5 cm. The arm of the short muscle is three times less, that is, 5/3 cm. By how many times more should the

---

[2] 16.38 kg (36.11 pounds).

[3] Brachialis.

short muscle get tense than biceps? (*The student finds it difficult to answer.*) Think of another number. Here is a load of 1 kg. What will be the tension of each of the muscles if they work separately? Let's say that the tension of biceps is approximately 7 kg. The tension of the internal muscle is 21 kg, that is, three times more than the tension of biceps. So we can see that a person who can hold a *pood* with his/her outstretched arm can tense his/her biceps muscle approximately up to 100 kg. That means that the maximum force is 100 kg. If the internal arm muscle is capable of holding the same load with an outstretched arm (in reality it cannot, but if it could), what amount of tension would be needed for this? If biceps is 100 kg, the internal arm muscle will be?

Student: 300.

Lecturer: 300 kg. Now imagine that we have an opportunity of some kind to make an artificial muscle. Let's take as bones two wooden planks and connect them with a hinge and let's stretch an artificial muscle between them, which is to represent biceps. We will build it in such a way that this muscle will be able to balance a load of one *pood* at the end of the wooden plank, that is, it will be able to develop a force of 100 kg. Now I suggest structuring the muscle so that it could produce the tension required to support one *pood*, if it is positioned like an internal arm muscle. What length and size should it be in order to hold up 300 kg? How much thicker should it be?

Student: Three times.

Lecturer: It's clear that it should be three times thicker. And how much shorter?

Student: Three times.

Lecturer: Now what about the size? In which of these two muscles should the quantity of muscle substance be larger or smaller?

Student: More … less.

Lecturer: Why more?

Student: The same.

Lecturer: Of course it should be the same. Why? Because the work will be the same. Whichever muscle we take, the work will depend only on how much muscle substance we are using. Here we've calculated correctly that they will be equally capable of work if they have the same size. I recommended several books to you; in one of them, a very good book by I. M. Sechenov,[4] *Essay on Man's Working Movements*), there is a discussion of what we have just been talking about. But there it says that short and fat muscles are significantly different from long and thin muscles. We saw that they do in fact differ but only when we make them lift a weight directly. However, if they are assembled to perform work according to the law of the lever, we will get a different picture. A long

---

[4] I. M. Sechenov (1829–1905) was a Russian physiologist, greatly revered by I. M. Pavlov and others who built on his work, and author of *Refleksy golovnogo mozga* (*Reflexes of the Brain*) in 1863. *Ocherk po rabochikh dvizhenii* (*Essay on Man's Working Movements*) was published in 1906.

and thin muscle can substitute for a short, fat muscle if we position them in a corresponding way. The appearance of fat muscles or thin muscles depends for the most part on what the general conditions of work of the muscle are, where it is situated and where it is convenient for it to be.

Now let's now consider how muscles serve whole joints with various degrees of mobility. As you remember, in the second lecture, it was said that a muscle in many respects resembles a cylinder of the internal combustion engine. By the way, this similarity is preserved in those two classes of muscles we have just considered. Any of you who has at some point dealt with cars knows the difference between the value of the diameter of a cylinder and that of the piston stroke. The role that the length of the lever arm plays for muscle is played by the gears and gearbox in a car. But there is a similarity also in the fact that the muscle and the piston of the car cylinder can produce active movement only in one direction and for the opposite direction they need an external force. Sometimes (in single-cylinder engines in motorcycles), a flywheel creates this outside force; in our experiments with the muscle in the second lecture we used a stretching load for the same purpose. However, it is much more convenient to create an engine out of two halves which act in opposite directions. In multi-cylinder internal combustion engines various cylinders help each other so that when one makes its active movement, it at the same time helps its comrades, which are joined with it via their common axle, to produce the passive part of the movement (to suck out the mixture and to compress it). Muscles are also organised in these artels. They are positioned on both sides of the joint in such a way that if one of them makes the joint turn in one particular direction, the rest, with active contractions, are able to turn the same part of the body back. Such groups of muscle of opposing action bear a special name of muscle antagonists, but this name, as we will see maybe even in this course, is very inappropriate. For biomechanics, it is essential that these muscles are not opponents among themselves, but on the other hand that as an artel they take part together in a collective way in carrying out a particular movement. However, it will only be possible to understand how the members of such artels are situated only after we have taken a look at the role of muscles from an entirely new point of view.

Up until now we have been constantly and persistently looking at muscles exclusively as engines. However, the main role (and maybe the most ancient one) of the muscles is to be an integral structural part of the body of an animal, which is vital for its stability. Take a look at the sentry who stands like a stone in front of some monument or palace. Where is the movement here? He is as still as a doll, but at the same time tries to place a bare bony skeleton in the same pose. We can see all the same hard parts as in a sentry who is alive; however without the whole set of supports the skeleton will not stand: it will collapse. The main difference is that the living person will need muscles in standing motionless as much as the bones. Muscles are in essence an inevitable static element of the human frame.

Here is another example. Muscles are soft tension braces, as we have already said. The ropes of a hanging bridge are the same sort of non-rigid tension braces.

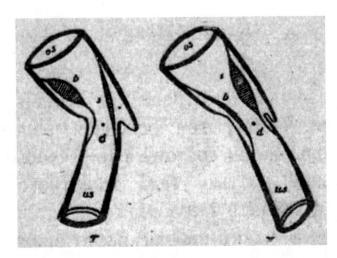

Fig. 4.3 Section of an insect joint and the positioning of muscles that operate this joint (by Shenikhen)

Try to cut these ropes, and you won't diminish a single hard part of the bridge. However, the cars and trains that have to use the bridge won't like it; they will have to have a swim in the river. Everybody knows this about the ropes, and so if you attempt to cut them, you will be arrested. But for some reason no one thinks in a similar way about muscles.

Let's compare the structure of the skeleton and muscles of a human being with that of insects. In insects, muscle only plays the role of movement and a dead, dried-out insect is as stable as a living one.[5] The segments of the limbs of insects have the structure of tough tubes that are of equal strength whether they are being stretched or compressed. Therefore, their muscles only play a role in movement (Fig. 4.3). They are placed inside the tubes and are occupied only with various pulls of the tubes.

It is different with vertebrates. Their skeleton is built everywhere according to one principle: the tough core in the middle and springy tension braces on the sides. The first works to compress, and the latter to extend. Such a principle, for example, is used to build radio and telegraph masts, which have a core in the middle and are pulled by cables on four sides.

The whole difference between artificial constructions of such kinds and the human machine is that the radio mast tension braces have a constant length and are almost not stretchable whereas in a vertebrate part of these tension braces can change their length and the degree of tension. Muscles in essence are simply cables, only cables that can regulate their length.

In the human machine of vertebrates, two types of tissue undertake resistance to compression; these are bony and cartilage tissue. These two types of tissue are

---

[5] In fact, insect joints do not hold firm without muscles although they dry out and become stiff after an insect is killed.

closely related, can be substituted for one another and often are developed from a common original source. In exactly the same way, the task of opposing the stretch is taken on by two other groups of tissues, which are also related and transform one into another. They are tendon and muscle tissue. The difference between these two, essentially, is only that tendon tissue does not regulate its length and muscle does.

I told you in the first lecture about a wonderful property of the living organism to eliminate bony tissue in the areas where it is not subjected to the direct influence of force. The organism has the same property in relation to other types of tissues; moreover, the tissue which has a finer and more complicated structure is always prone to be substituted by another one, less highly qualified, at the first opportunity, if its special abilities are not in demand. In particular, muscle tissue is constantly replaced by tendon tissue, that is, by ligaments in those areas where there is no demand for self-regulated tension braces. We know, for example, from medical practice what happens when some joint loses its mobility as a result of an old sprain or some illness. Inevitably, degeneration of muscles of this joint takes place after this, and they turn into tendon-like tissue. The positioning of muscle around the joint becomes clear only when we consider its ligaments and muscles together as one whole.

The oldest forms of joints had a very indefinite form and pliancy in many degrees in all directions. Because of this, the muscle covering of such joints surrounded them more or less evenly on all sides. We have to consider the transformation of separate parts of this muscle cover in tendons and ligaments as a secondary change, which is connected with the more exact formation of joints and the limitation of their degree of mobility, which comes with this.

Take, to begin with, a lever that is put onto an axis with one degree of freedom. Such a lever can turn only on one plane; as regards all other directions, it is fixed by the form of its hinge. That means that it requires only as many tension braces as are necessary to fix it in that direction for balance. So how many tension braces will it need?

Students: One? Two?

Lecturer: Look at Fig. 4.4 and try to understand the positions in it. If we limit ourselves to only one tension brace, it will pull the lever on its side and nothing will interfere with this. It is obvious that there should be two tension braces that, moreover, are capable of the directly opposite form of actions. The joint is fixed fully with such tension braces.

Now let us have a look at the two-axis joint (Fig. 4.5). How many tension braces do we need if we have two axes? People often think that for a two-axis joint two pairs of tension braces of opposing action are needed and three pairs are required for a three-axis joint but this is not correct, however. As we can see from Fig. 4.5, three tension braces are enough when you have two degrees of freedom; moreover they can be positioned randomly so long as they are not all three on one side. You understand that if we position tension braces in the form of a perfect triangle (as it is done, for example, with tents that have three cables) then by changing as appropriate the length of these three tension braces, one can make the axial pivot take any position

Fig. 4.4 Sketch of muscle
tension braces of a
one-axis joint (muscle
antagonists)

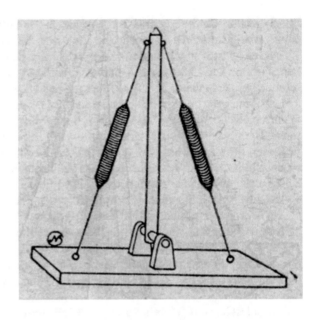

Fig. 4.5 Sketch of the
arrangement of muscle
tension braces of a
two-axis joint

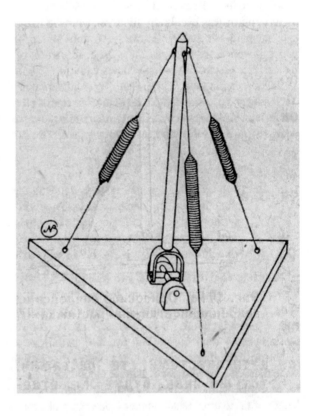

out of the number of positions that it can have, according to two degrees of freedom. In exactly the same way, one could prove that with a three-axis joint, the minimum and sufficient number of tension braces is four. Obviously, in the last two cases, a single muscle tension brace is not an antagonist to another stable opponent. Quite the opposite, a bony lever can be balanced in any position possible for it only by means of the corresponding redistribution of the lengths of all its tension braces. It is not just one muscle with the appropriate direction of its fibres that takes part in each movement in the joint, as people used to think, and not two muscles of the opposite action as they sometimes think even now, but it is the whole body of muscles that are connected to this given joint. Each movement of the living organism is a result of the redistribution of tension between all the muscles of the organ that produces this movement. Moreover, it happens quite often that various parts of the same muscle simultaneously have different tensions, so that the process of redistribution is spread not only through separate muscles but also through their constituent parts.

Imagine that a joint that used to be mobile in all directions gradually acquired a limitation on its mobility. It developed a clearly formed apparatus of articular endings with flanges that prevent all-round mobility. Because of this, it can no longer perform many of the movements in the range that it had before. Now you can see that muscles that, thanks to this, are doomed to be stationary do not remain muscles but turn into ligaments. We've established that for a joint with the greatest three degrees of freedom we need only four tension braces; we will look for these four braces now in any joint; only remember that the muscle construction has preserved from them only the quantity that is necessary for the joint with the given mobility.

In other words we will expect to have two muscle braces (of opposing action) in a one-degree joint and two ligaments; in a two-degree joint, three muscles and one ligament; and in a three-degree joint four muscles and maybe not a single ligament. In many cases we will find such a distribution in life.

Now we have in our hands a key to working out the arrangement of muscles of the human machine. We have an opportunity to approach their distribution consciously and with exact data in our hands. The description of the distribution itself and of the way the muscles of the human body work is the subject of the next lecture and now I want to teach you how to define the function and the direction of the action of the muscle in the living body.

You remember that the muscle gets tense when there is direct evidence of some kind of stretching force applied to it; remember that action is always equal to counteraction. If no external force prevents the muscle from completing a movement, then it will make it comfortably; it will contract and this does not affect it with any tension, or in general with any obvious indication; it will remain as soft during its work as it was while at rest. So if we go this way we won't learn anything.

That means that in order for the muscle that is designated to perform a given movement to give itself away in some way, it is necessary to get it to become tense and hard to the touch. For the tension, one needs a counteracting force. The method is as follows.

For instance you want to define which muscle extends the shoulder. You ask a volunteer to extend the shoulder; but even before he starts this movement you take hold of the upper arm and with all your strength prevent him from making this movement. He is trying to extend the shoulder and you don't let him. In such conditions all his muscles will remain soft, apart from those that in usual conditions produce extension of the shoulder. If you keep holding his arm with one hand and with another touch lightly with your fingers various muscle and muscle bundles around the shoulder through the skin, then you will immediately catch those that are involved in this particular movement. This is the only correct way, which I recommend you to use in all cases.

Let us do a test. I will ask three people to come out here to the front of the stage. One will be the volunteer and two others will be investigators.

Try to determine which muscle takes part in extending the upper arm. At first make this movement: Do you remember what extending the upper arm is? True, it is raising it forward and outward.[6] Now experiment in the way I explained to you. Hold his arm as strongly as you possibly can; and he will do all he possibly can to extend the upper arm. Feel his shoulder area. Can you tell where the tension is?

Students:	In front, on the shoulder.
Lecturer:	Take chalk. Let one of you continue holding his arm and the other will examine carefully the borders of tension of muscles and will draw them with chalk on the skin. What has happened? You see that as the result of the experiment a muscle has defined itself, sitting on the outer front side of the shoulder area, like an epaulette. This muscle is called deltoid; and now you know how it works. You can be sure after doing this experiment that the extension of the arm that is not loaded will be done by this very muscle (Fig. 4.6Б). Let us make another experiment. I will ask our volunteer to raise the upper arm by 45 degrees straight out to the side. No, you are not doing what I am asking. You are raising the shoulder girdle but the upper arm is the section between the shoulder joint and elbow joint. That is correct now. Do you remember how to adduct the shoulder? (*The volunteer lowers his arm back down.*) No, the movement you've just made will simply have to be called lowering the upper arm. Adducting is a turn of the upper arm about its vertical axis. Put your upper arm into its previous position at 45 degrees to the side. Now make a turn with it in front, and keep the elbow at the same height all the time, taking it along the horizontal circle. That is it. That is adduction of the shoulder. Now please come here, investigators, and find out which muscles make this movement. Don't get lost. What do we need to do here?
Students:	Hold the upper arm?

---

[6] As in the previous note, this would now be considered as shoulder flexion.

Fig. 4.6 Muscles of the right shoulder region, from the front. (A) Pectoralis major. (Б) Deltoid (by Spalteholz)

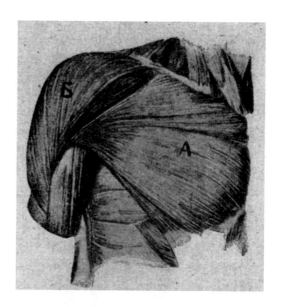

Lecturer:	Exactly and how will you do that? That's right: do not let him adduct the upper arm. Wait a second before you feel it; turn the volunteer to face the auditorium directly and do it this way: let him in turn try to adduct his shoulder strongly and then relax his arm back to the original state. We will ask the audience; can they tell us, without feeling it, something about which muscle is interested in the adduction of the shoulder?
Students:	The area under his armpit is getting tense.
Lecturer:	Only under the armpit? Look closer.
Students:	On his chest as well.
Investigator:	He has a band tensed in front of the armpit and also the whole front part of the chest.
Lecturer:	I will show you a drawing that depicts the muscles of the shoulder region and the ribcage in front. Will you recognise on it the muscle that you find guilty of adducting the shoulder (Fig. 4.6)?
Investigators:	This one. (*They point out the muscle indicated in Fig. 4.6 by the letter A.*)
Lecturer:	Yes, you have found the muscle that is in demand correctly, again. This fanlike powerful muscle is called pectoralis major; we will get acquainted next time with the ways it works, which are very varied and with the places where it is attached. Now you are armed with a method that will help you to orientate yourselves with the action of any superficial muscle. Train yourself in your spare time.

Lecture 5

Lecture 5 continues the investigation of the workings of muscles in relation to the upper limb.

Comrades! We are starting an investigation of the muscle and ligament equipment of the human machine. I'm not going to bother you with descriptions of individual muscles. After all, last time we clarified that the role of the muscles in movement and statically does not fit with anatomical divisions of separate muscles. What anatomists name as separate independent muscles, they often single out from the general muscle mass on the grounds of such incidental features as a somewhat denser capsule of connective tissue covering some part. By freeing ourselves from the obligation of speaking about anatomical details, we can make our task much simpler and reduce the time needed for explanation. We will start with the torso.

I have already mentioned that the spine of a four-legged creature as a structure represents a bridge slung between two buttresses—the fore- and hindlimbs. Look at Fig. 5.1. The spine between the two points of support makes an arch, with the convexity, as you would expect, at the top. If only rigidity was required in the spine, then just this curve would be sufficient to strengthen it. But nature has given the spine a more complex construction requirement, a requirement which, one must say, has not yet been resolved by the art of construction of humans. Namely, the spine, while preserving its firmness and not bending or breaking, has to be a very flexible structure at the same time, to have the opportunity to change its form in various ways. The solution to this situation found by nature is as follows.

The principle for structuring of the spine is the same one we have already encountered during this course more than once: nature is mean with its inventions. The firm rod of the spine is supplied with tension braces on four sides (Fig. 5.1), which go in parallel with it along its length: on top, down below and on both sides. These tension braces have variations in structure. You understand that a constant demand for the resistance to being stretched is imposed on the lower tension brace; it acts all the time while the animal is standing. In connection with this, the lower tension brace, a tendinous one, is structured like a flat ribbon, which is stitched to the bodies of all

© Springer Nature Switzerland AG 2020
N. A. Bernstein, *Biomechanics for Instructors*,
https://doi.org/10.1007/978-3-030-36163-1_6

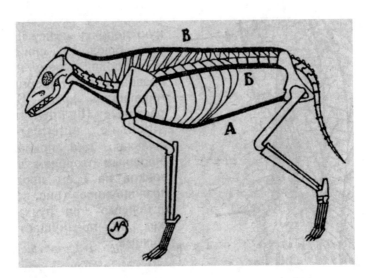

Fig. 5.1 Scheme of the muscular tension braces of the four-legged mammal. *A*—ventral tension brace, *Б*—tendinous tension brace of the spine, *B*—spinal tension brace

vertebrae from the ventral (that is, lower) side. The other three tension braces, which sometimes work and sometimes don't, at any rate, vary in varied situations and have a muscular structure. And the back one (spinal) and both side tension braces lie in the form of two fat straps on both sides of the spine. This is the 'meat', which is known in kitchens as chops. The whole mass of muscle fibres of these tension braces goes longitudinally; for the most part, the fibres are mainly short here, and are stretched between vertebrae that are next to each other and those that are close by. They not only connect all vertebrae, the neck, thoracic, lumbar and sacral, but also capture all the bones that turn up in the neighbourhood: the occipital part of the skull, back ends of ribs and pelvic bones. Look once again at Fig. 5.1; you see that the spine of the four-legged creature has three different curves along its length. In the thoracic part, its convexity faces upwards, in the neck and lumbar area, and we have the concavity upwards. Think where you need the firmest tension braces, especially at the top. After all, the spine in the thoracic area looks like an arch; that means that its firmness is achieved, to a large extent, by its form. It is not the same as the cervical and lumbar areas. Indeed, in these areas, the longitudinal muscles of the back are fatter and more solid than in the thoracic part.

As regards a human being, with a spine arranged vertically, the distribution of forces is different. You remember what we said in our previous lecture about the moment of force. The smaller the arm of the lever, the stronger the force needed so that balance takes place. The force in this particular case is the toppling force of gravity, which tries to make the spine fall downwards. For a given value for the moment of this force, the action forces will be stronger, the closer they are to the bottom end of the spine. That means that in the lower part of it, the longitudinal muscles in a human must be fatter than in the upper part of the spine. Do you understand?

Students: Not entirely.

Lecturer: Think: the top sections of those muscles, for example, the neck parts, have to sustain the load only of the head, which weighs 4¼ kg, and the arm of the lever here from the middle of the cervical section to the centre of gravity of the head is about 15 cm. The lumbar area, however, has to resist the action of gravity of the torso, with the head and arms, which weigh 38 kg, and the arm of the lever here is about 35–40 cm. That means that for the lumbar muscles the load is nine times more and the lever arm is three times as big. That means that the moment here is 27 times more than for neck muscles. Do you see now? Right then.

Now we have equipped the arch of the spine. We can go on to have a look at how the spine is loaded. The torso is the load of the spine. In a four-legged creature both pairs of limbs serve as supports and they are not taken into account: only the torso itself hangs on the spine, presenting itself as a hollow box or suitcase for the internal organs. All the muscles, bones and ligaments of the trunk (apart from a few muscles of internal organs which we are not going to mention here) represent just the equipment of the walls of this suitcase. And the construction of these walls is exactly the same as that of many man-made suitcases.

These walls are wrapped in three layers of flat muscles with fibres that go in different directions. In the two outer layers, the fibres are laid out diagonally and criss-cross in relation to each other. In the third, deepest layer, they go horizontally and vertically. Thanks to such a cross-layered arrangement the walls are firm and elastic in all directions. If you like, sheets of veneer are made using the same criss-cross-layered principle; you know how beneficial this is for firmness. In the abdominal part, the muscle layers go throughout; in the thoracic part you have ribs, too, going through them, which don't change anything in the way they are arranged, or in the way these layers act. Biomechanic people have a very bad habit of comparing everything they can with artificial constructions. In accordance with this habit, I cannot help comparing ribs here with whalebones in a corset.

I can't give more details here in describing muscles of the walls of the trunk; I suggest that those who are interested have a look at an atlas of anatomy.

The much more sophisticated machines of the girdles of the limbs are mounted on this simple construction of the trunk. These machines can only be understood if we look at the four-legged mammals first. We have already mentioned that the shoulder girdle is built using the principle of a hanging bridge. It is fixed to the trunk in four-legged creatures by a width of muscle material on which the rib cage is placed, in the same way as they put someone who is learning to swim into straps. This sling of material captures, in consequence, the whole rib cage from the bottom part; and at the top it grows into the top ends of the shoulder blades and goes higher and both parts meet at the spinal column. In a human being, the proportions change. It's not his/her front limbs that carry the body, but the opposite: the trunk carries the front limbs. That loads the shoulder girdle three times more. Apart from this, in the vertical torso the muscle bandage that I just described lies horizontally and not vertically as before and serves as a hanger for the trunk, only indirectly. In connection

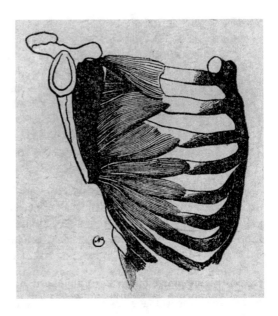

Fig. 5.2 Front part of the muscle bandage which fixes shoulder blade to the trunk. Serratus anterior (by Spalteholz)

to all this, it has become weaker and smaller and other parts have strengthened correspondingly on account of this. A view of this muscle bandage from the front in a human is depicted in Fig. 5.2 and we mentioned its back half, the one that connects the shoulder blade with the spine, in lecture 4, on p. 57.

And so the human being has developed new secondary tension braces, which are useful for hanging the shoulder blade and the shoulder girdle, in the new vertical stance. These new hangers support the shoulder blade from the top and from below, strengthening it in those directions towards the spinal column. You can see what these hangers, the trapezius muscles, look like in Fig. 5.3.

Thus, the human shoulder blade is supplied with four muscles in one complex, which go from it in all four directions to the wall of the torso. I have already mentioned that four tension braces can provide three degrees of freedom; and these three degrees are exactly what the shoulder blade has. It can make movements up and down and to both sides and besides this it can also rotate. From the previous explanations it is clear that the sideways movements of the shoulder blade (adduction and abduction) can be performed with the help of both halves of the muscle bandage, which serve as the front hangers in four-legged creatures. However, its up and down movements are carried out by the parts of the new hanger, the trapezius muscle. At the same time, the top, lifting part of the trapezius muscle is stronger and bigger than the lower one. There's nothing strange about this: after all, it has to overcome constantly the force of gravity of the arms.

What happens with the rotation of the shoulder blade? The shoulder blade muscles have to contract in portions, in parts. To turn the shoulder blade outward we have to use the lower portion of the spinal half and upper portion of the ventral half of the transverse tension brace. To turn the shoulder blade inwards there are two

Fig. 5.3 Muscles of the
back. M = trapezius;
Ⅲ = latissimus dorsi (see
Fig. 5.6); Д = deltoid (by
Spalteholz)

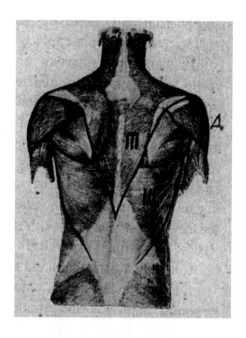

whole muscle mechanisms. Firstly, the very same transverse tension brace can turn inwards too, if only the upper portion of its spinal part and the lower portion of the ventral part contract in it. But besides this, the trapezius muscle can give the same effect. Look at how it is constructed (Fig. 5.3). With its torso end it is fixed to all the cervical and thoracic vertebrae, and even to some of the lumbar. From all this huge extent, it fans out towards the shoulder blade, where it is fixed to the big bony lever that protrudes far out of the plate of the shoulder blade. But the fibres of trapezius are not fixed to this lever only in one place. Quite the opposite, the uppermost fibres, which go from under the occiput, go increasingly outward and are fixed on that lever of the shoulder blade already mentioned at the outermost end, sometimes even going beyond the clavicle. And the lower fibres of the same muscle, on the contrary, ascend almost vertically and end at the shoulder blade at the innermost end of the same lever. Imagine that the top and lower portions of trapezius tense at the same time. It is clear that such tension must lead, again, to turning the shoulder blade inwards.

How can we explain the fact that to turn inwards there is a mechanism that is twice more powerful than for turning it outward? If you think only about the shoulder blade you won't be able to explain this. I will use the opportunity here to check what you have taken in regarding the mutual movements of the shoulder blade and the upper arm. Imagine that the upper arm is fixed motionless in the shoulder joint with the help of muscles from this joint. What will happen with the shoulder in this position if you start turning the shoulder blade inwards? Is it difficult for you? I'll explain in other words. Try to figure out which movement of the upper arm is helped by the turning of the shoulder blade inwards. I can say this again in other words. You remember that movements of the shoulder blade can increase the limits of the mobility of the shoulder. So during which movement of the shoulder do you have to start the turn of the shoulder blade inwards in order to extend the limits of the mobility in this direction?

Students: During shoulder extension.[1]
Lecturer: That's right. And consequently, how will the shoulder blade accompany the opposite movement—shoulder flexion?
Students: With rotation in the opposite direction.
Lecturer: Yes, that is, rotation outward. Now tell me, what do you expend more force on: extension or flexion of the shoulder?
Students: On extension.
Lecturer: Why?
Students: Because there is a lift upwards when extending.
Lecturer: That is why the rotation of the shoulder blade inwards requires a stronger apparatus of muscles than for rotating it outwards. In essence, that is all there is to the whole suspension of the shoulder girdle of the human being; but not all muscles of this area have been considered. The reason for this is as follows.

When the development of the muscles of the limbs begins, these muscles appear from their lower ends at first, that is, on the limbs, and from there they stretch towards the girdles and the trunk. Some of these muscles are able to crawl only to the bones of the girdle areas—in our case just to the shoulder blade—others go right through without any stop through the shoulder blade and are fixed to the torso itself. It is similar to how suburban trains work on many railways; some go only to nearby stations, and others only to the remote ones. So these non-stop trains for distant stations, that is, the muscles of direct 'shoulder-torso' communication, mix with the suspending muscles which we have just described and complicate the external picture and also, partly, their mechanical behaviour.

It is easier for us to start with the short muscles of the shoulder joint. There are five of them around the joint; it is more correct from the mechanical point of view to consider one of these, that is, deltoid, together with the muscles of the long group. The remaining four are as follows (Fig. 5.4).

They all join the upper arm with the shoulder blade. They approach it from both sides in pairs. On the outer side of the humerus at its very end there is a small bony protrusion, which is called the greater tubercle of the humerus. At the front edge of the humerus, right at the top, is another protuberance—the lesser tubercle of the humerus. Two of the muscles mentioned go directly to the greater tubercle, one from above and the other from behind. In Fig. 5.4 these muscles are indicated by numbers 1 and 2. The one that approaches the tubercle from the top starts on the shoulder blade, just above the bony lever that I mentioned when describing trapezius. The second one starts under the same lever, from the whole of the shoulder blade plate. If you look at their positioning carefully you will see that the top muscle

---

[1] See previous notes on shoulder flexion and extension.

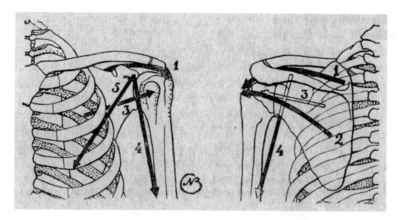

Fig. 5.4 Direction of the muscle pulls of the short group of the shoulder: (*1*) Supraspinatus, (*2*) infraspinatus, (*3*) subscapularis, (*4*) coracobrachialis, (*5*) pectoralis minor

can work as the extensor of the shoulder and the lower muscle as the flexor. Apart from this the lower muscle, which captures the humerus from behind, can be used also for rotating the shoulder outward, around its long axis.

The third muscle in this group starts also from the whole area of the shoulder-blade plate but this time from its inner surface, the surface which adjoins the ribs. It goes towards the lesser tubercle of the humerus and skirts around the shoulder in front. That means it can make the humerus rotate inwards and apart from this, con-tracting together with the second muscle that has just been described, produces flexion of the shoulder (Fig. 5.4, No. 3).

And finally, the fourth muscle of this group starts from a small hook on the shoul-der blade near the shoulder and goes down along the shoulder and grows into its inner edge. This muscle is clearly an adducting muscle of the shoulder. You can ask why there are two muscle groups in all in the shoulder joint, long and short, and what duties are distributed between both groups.

Maybe an experiment will answer our question best. Let us organise it like last time: with one volunteer to demonstrate and two investigators. Last time we didn't look at the muscles of the hand, so let us look at them now. The volunteer will make a fist and one of you will prevent this movement, in the way that you know. I will ask the second investigator which muscles take part in squeezing the fist. Do you remember how to do that? Yes, of course, you have to touch the muscles, so do that.

(*One person makes a fist: their fingers are held by another person; and the third feels their forearm, upper arm and shoulder area and then points at the front of the shoulder and says: this is where the tension is.*)

Lecturer:	And where else there is tension?
Investigator:	It is also up here.
Lecturer:	Does this mean, in your opinion, that muscles of the fingers are situ-ated in the shoulder area?

Investigator:	So it seems.
Lecturer:	Try the forearm as well.
Investigator:	It is tense here too.
Lecturer:	What does this all mean then? Is it that muscles of the whole hand or arm get tensed to make a fist? A little further on, you will see that the real flexors of the fingers are situated in the forearm, that is, where you found them. So what role is played by the other muscles, which are situated higher? Let's change our experiment. The volunteer will remain as he is and will try to form a fist in vain again, and you at the same time investigate the area of the shoulder joint from all sides, from the front, back and top.
Investigator:	The tension is everywhere; however, it's a bit less here and here there is more.
Lecturer:	Let us now consider only two of any opposing directions, for example, the front and back of the joint. Do you think that the muscles that are positioned on the opposite sides can work together or they are in essence antagonists, so if they tense there will be consequences that are directly opposite?
Students:	Perhaps they will work against each other.
Lecturer:	Well this is the thing: what can we get when they work jointly apart from immobility? They can only brake each other. That means that you can see here a new way in which muscles work, not the same way we researched before. Now what do you think: what can this simultaneous tensing of opposing muscles be used for?
Students:	To immobilise the joint.
Lecturer:	Exactly. Now maybe you will be able to work out also how we found tensed muscles in the shoulder region when the fingers were moving. The fact is that our experiment was not conducted properly. If an investigator is trying to open the fist of the volunteer, then he pulls his fingers with force, and the whole arm is pulled along with the fingers. The finger muscles cannot undertake anything in order not to the make the whole arm move; therefore, the muscles of the shoulder and the shoulder girdle get used. Consequently, the hand can rest when the shoulder works; but the shoulder has to work when the hand works.

That's not all. Let's come back to calculating moments. Let's ask the volunteer to stretch his arm forward and I will place this 1 kg scale weight on his palm. Does anyone have a measuring tape? Let's measure the distance from our weight to the centres of all the arm joints. Look: from the centre of the weight to the radiocarpal joint it is 9 cm; from the same place to the elbow joint it is 34 cm; and finally, it is 65 cm from the same centre of the weight to the shoulder joint. That means that the moment of our weight in relation to the elbow is almost four times more than in relation to the wrist; and in relation to the shoulder it is seven times more. Thus, even if the arm didn't weigh anything and it had been only the scale weight itself, then the load in the shoulder area would have been seven times greater than on the hand even in that case. Not only do the muscles of the shoulder get loaded when the hand is working but they also get loaded several times more than the muscles of the hand itself.

This is how we can explain the fact that the shoulder has to have reinforced, double-muscle equipment. It works not only on its own behalf, but also for the whole arm, and in the latter case it requires even more force than in the former. Therefore, the shoulder muscles have to perform two types of duties. Firstly, they move and turn the shoulder in all possible directions and secondly, they secure, that is, fix the shoulder in every required direction. In very many cases, this second duty of fixing is performed by the muscles of the long group that we have not yet described. The muscles of the short group, which are weaker but quicker moving, are usually sufficient for turns and shifts of the shoulder. Maybe the passage of the long muscles from the bones of the shoulder area onto the trunk can be explained precisely by the need to get a stronger and wider area of support.

Now we can go on to scrutinising the long group of upper arm muscles. From the anatomical point of view, this has three muscles but from the biomechanical point of view it has at least five. This happens because the deltoid muscle is very wide; it captures the shoulder area from three sides and its outermost bundles of fibres have entirely different and almost opposite action. It is depicted with the help of three arrows, 1, 2 and 3, in Fig. 5.5.

Maybe it's not even worth describing how these three different bundles act. If we look at the drawing attentively this action is depicted very clearly. Besides, deltoid

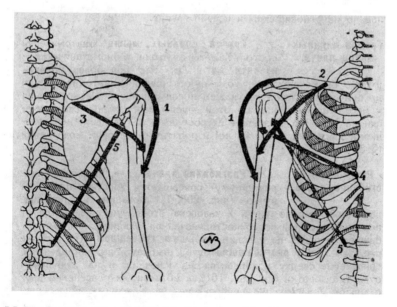

Fig. 5.5 The directions of muscle pulls of the long group of shoulder muscles. (*1, 2, 3*)—separate bundles of deltoid, (*4*)—pectoralis major and (*5*)—latissimus dorsi

lies very superficially and you can check how it contracts yourselves very easily. Obviously, bundle No. 1 is the real extensor of the shoulder (in this it is similar to No. 1 of the short group). Bundle Nos. 2 and 3 of the long group act as a flexor when they work together, but when they work separately they can perform adduction, abduction and rotation of the shoulder.

We need to consider bundles 4 and 5 in more detail. Both are shown in the drawing simply with arrows. But this is just a convention. In reality, both are very big, wide muscles. Both cover a very wide surface on the torso and then draw together in the form of a fan towards one small tendon and there grow into the shoulder bone. Bundle No. 4 starts on the entire front surface of the rib cage. It is pectoralis major, which is very visible in muscular people. If you remember, last time we proved that this muscle is responsible for the adduction of the shoulder. If you try to perform our usual test for adducting the shoulder, this muscle, together with its tendon, will contract and will be quite visible under the skin. As you can see, pectoralis major creates the front wall of the armpit.

The fifth bundle of the same group is the widest of all the body muscles. It is shown separately in Fig. 5.6. There you can see that it starts very low, from the pelvic bones and the sacrum and also from the lumbar vertebrae. This entire wide sheet

Fig. 5.6 The position of latissimus dorsi on both sides and the way they attach to the humeri. All other back muscles are removed for clarity (by Mollier)

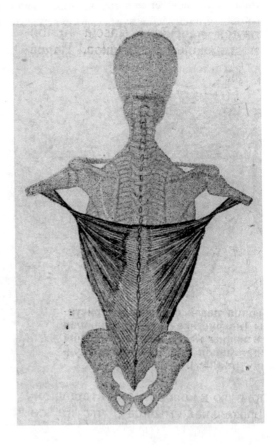

bends round the back from the bottom upwards and ends with a very narrow tendon under the lesser tubercle of the humerus (that is, in front). From this drawing it is easy to understand how this muscle should function.

We must add something to this. All muscles of the long group should differ from the short muscles by the fact that they terminate on the bones of the torso, missing the shoulder blade and the clavicle. However, this is not the case for deltoid; and you will remember that at the beginning of the lecture I referred it anatomically to the short muscles. In reality, we should understand it as follows: this muscle has only a break, a stop on the shoulder blade and clavicle; its continuation upwards is nothing other than the top part of trapezius, which goes as far as the vertebrae.

After such a long account of the muscles of the shoulder joint, the description of the elbow muscles will seem a trifle to you. The more mobile the joint is the more muscles it has; and the elbow joint is regarded as one of the simplest, one-axis joints. According to the theory, we should expect to see here two muscles and two tendinous cords. The muscles must be situated on both sides of the axis and tendinous cords at the ends of the axis.

And that is how it is. The elbow joint (that is, the joint of the humerus and ulna) has one flexor and one extensor. The flexor is constructed very simply: it starts approximately in the middle of the humerus and ends at the very top end of the ulna, on that prominent point, which represents the front flange of its joint cavity at the same time. The extensor of the elbow is constructed in a more complex way.

It is fixed to the ulna absolutely symmetrically to the flexor and on the same sort of prominent point as the former. Only it is much easier to feel this prominent point because it is situated on the convex side of the elbow. This is the point of the elbow. The low end of the elbow extensor bulges under the fingers when you try to extend the elbow but at the same time restrain your arm.

The top end of the extensor is divided into three heads, of which only one ends on the humerus; the other two go as far as the shoulder blades; that is, they cross both joints. The extensor of the elbow (triceps) is the only muscle at the back of the upper arm; it performs elbow extension and can help to flex the upper arm in the shoulder joint.

That would have been the end of the story, if we hadn't had the radius next to the ulna bone. This bone has its own special muscle to flex it. This muscle is best known to the wider public, for some reason, maybe because of its superficial position. It is called the biceps or two-headed muscle of the upper arm.

However, because of the way it functions, biceps is one of the most complicated muscles. To begin with, it is not attached to either the humerus or the ulna. It starts on the shoulder blade and extends from there by a straight route to the radial bone. Therefore, it crosses three whole joints (shoulder, humero-ulnar joint and radio-ulnar) and it can bring about movements in all these joints. Imagine a chain made of four little segments, and between the ends of the two extreme ones, an elastic band is stretched. The conditions of reciprocal mobility of these little segments are completely different. Can you predict for certain how this elastic band will change the form of the chain when it contracts? Clearly, one cannot do that. There can be some kind of certainty only if we fix two joints out of the three in this chain with something else.

Moreover, the certainty will be different every time depending on which two joints out of the three we secure.

The very same thing happens to biceps. This muscle, without the help of the others that fix the joints that are superfluous for it, is entirely useless. It is utterly impossible to describe its action when it is isolated. One can only tell what it will do if we fix two joints out of those three, through which it passes, and leave the third one free.

In the shoulder joint, biceps helps with extension[2] and partly with adduction. In the elbow joint, it acts as a flexor. Finally, in the radio-ulnar joint it acts as a strong supinator. It is not by accident that nuts and bolts are constructed in such a way that screwing them requires strong force and is done with the help of supination (the so-called right-hand thread). The mechanism of this latter movement is as follows: the lower end of the biceps is fixed to the radius with a long tendon, which looks like a tape. When moving—in pronation—this tendon winds round the radius as if round a shaft. To perform supination, it uses force to untwist the radial bone back, as the string that is wound around it unwinds a spinning top.

It's clear that the greater the force of a muscle, the greater its moment in relation to the given joint. And the greater the moment, the greater the lever arm, that is, the distance from the point of application of force to the centre of the joint. Biceps goes the furthest from the centre of the elbow joint, which means that this muscle is the strongest in the role of the flexor of the elbow. For extending the shoulder and supinating the forearm, it is four or five times weaker.

Apart from the three muscles that we have just described, the elbow joint has two tendinous tension braces, which are situated on both sides of the joint. These tension braces secure the joint bones at the ends of the joint axis and apart from this they stretch along the whole length of the humerus like partitions between the flexor and extensor. This example shows especially clearly how the original continuous muscle sleeve turned into tendon in the places where it could not contract.

We could have expected a new complex system of muscles at the forearm area in accordance with those three degrees of mobility that the wrist has in relation to the elbow. Don't be scared: there is no other 'shoulder joint' in the whole human machine, and we have already moved on from the most difficult. The muscles on the forearm are sketched out very simply.

Firstly, it has three very short muscles for pronation and supination: two pronators and one supinator. This is natural because biceps helps in supination very powerfully. There are only four other muscles that control the radiocarpal joint and they are situated in a very straightforward way.

Just remember one thing. There are two bony protrusions at the lower end of the humerus on both sides of the elbow joint. They are called condyles: outer and inner. The thing you have to remember is that all flexors of the lower part of the arm start on the side of the inner condyle and all extensors from the side of the outer condyle.

---

[2] See previous note on Bernstein's use of 'flexion' and 'extension' terminology.

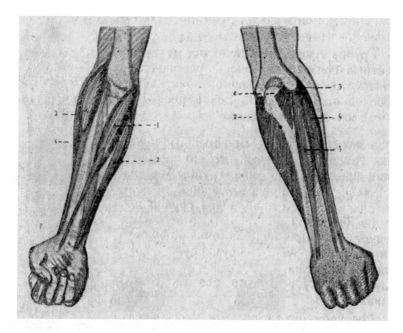

Fig. 5.7 Muscles that are responsible for the movement of the wrist (hand). Left side—view of the right forearm from the front. Right side—the same view from the back. (*1*)—radial flexor. (*2*)—ulnar flexor (*3–4*)—radial extensor. (*5*)—ulnar extensor (by Mollier)

As far as the hand is concerned, we have to distinguish the radial and ulnar sides. The radial edge, like the radius, is located on the side of the thumb, and the ulnar side and ulnar bone on the side of the little finger. So the four muscles that move the hand can be divided into pairs in two ways; firstly, two of them are in essence flexors of the hand and start from the inner condyle and the other two are extensors and start from the outer condyle. Secondly, two of these muscles stretch to the radial edge of the hand and two to the ulnar edge.

Thus the four muscles are as follows (indicated in Fig. 5.7):

Ulnar flexor of the hand	2
Radial flexor of the hand	1
Ulnar extensor of the hand	5
Radial extensor of the hand	3–4

All this is depicted schematically in Fig. 5.7.

Now think, what will happen if both flexors contract, that is, No. 2 and No. 1?

Students: There will be flexing of the hand.
Lecturer: And what if both extensors contract, that is, No. 5 and Nos. 3–4?
Students: Extension of the hand.
Lecturer: That is clear. Now what if both elbow muscles contract, that is, Nos. 2 and 5, and both radial muscles, 1 and 3–4, stretch? Do you find the

question difficult? What will happen if the ulnar side of the hand goes towards the forearm and the radial edge, the opposite, goes away from the forearm? You are making the movement absolutely correctly; what is it called?

Students: Adduction of the wrist?

Lecturer: Good; therefore if the same muscles are contracting in the opposite way then we will have abduction of the wrist. Is that clear?

Students: Yes.

Lecturer: Thus the radiocarpal joint is served entirely by our four muscles. All I have to do now is to show you where these muscles and their tendons are actually located. You all know that the forearm narrows down towards the lower end: this happens because at the lower end of the forearm there aren't any muscles at all, only very thin tendons. But the tendons of both extensors of the wrist can be seen easily at the base of the hand if you prevent someone from adducting it. They work in that case like two strong laces on the sides of the hand. The same thing applies to both flexors. The bellies of all these muscles are positioned obliquely because they start near the condyles of the humerus, which are situated on the sides. They can even be seen in a man with well-developed musculature, if we do the same experiment with restraining the movement of the limb.

The most interesting circumstance, and perhaps something that most of you do not know, is that almost all the muscles of the fingers also lie on the forearm in its upper end. That is very beneficial from the practical point of view. The wrist has only one tool—the fingers; the engines of the movement are located at a distance and, thanks to this, don't interfere with the mobility of the hand and don't slow down the mobility. Here, as well, the muscle bellies start near the middle of the forearm and further below the long tendinous laces stretch as far as the fingers, like transmission belts. The thumb is serviced in a special way and it's better to talk about its muscles separately; the other four fingers have muscles that are very closely connected to each other.

The muscles of the four long fingers consist of two flexors and one extensor (Fig. 5.8). They start at the forearm and partly at the lower end of the humerus together with the flexors and extensors of the hand on the corresponding condyles. The tendons of the extensor are stretched along the dorsal side of the hand right up to the fingers. You have all seen them on yourselves many times. The tendons of the flexors are more complex. The thing is that one flexor lies on the forearm deeper than the other one and each of them sends tendons to all four fingers. That means that every finger gets two laces, going one under another along the palm side. The laces of the superficial flexor end on the middle phalanges of the fingers. The laces of the deep one go as far as the phalanges where the nails are[3] and to do this they have to crawl their way under the corresponding tendons of the superficial flexor,

---

[3] By 'nail phalanx' Bernstein refers to what is generally known as the distal phalanx.

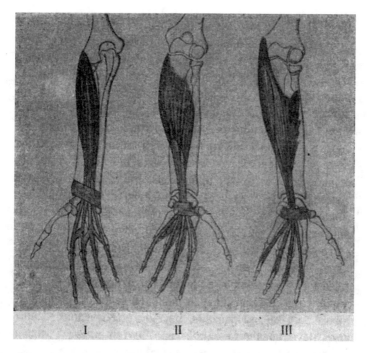

Fig. 5.8 The muscles which control the movements of the fingers. (I)—the extensor of the fingers, (II)—the deep flexor, (III)—the superficial flexor (by Mollier)

which are divided like a fork at the ends for this eventuality. But you should not think that the deep flexor flexes one nail phalanx alone and the superficial only one middle phalanx. Quite the opposite, you know that to flex individually only one nail phalanx without flexing the others is not possible for us. You have to remember that the muscle moves all the joints through which it passes. Consequently, the deep flexor of the fingers flexes the hand and all the joints of the fingers; that is, it forms the hand into a fist. The superficial flexor flexes all those joints, apart from the final joints of the fingers.

I am not going to describe to you in detail those muscles that are situated in the hand itself. These muscles are very small and weak and are only used in very fine work of the fingers, for example, in drawing or in playing musical instruments, and at that always together with the long muscles. Knowledge of the subtleties of defini-tion of these muscles will not give you anything; it is much more important to take note of one general principle. I have already said that the long muscles of the fingers go through very many joints. That means that what I said to you about biceps is applicable to them to an even greater degree. This could be stated in relation to the fingers in this way: the flexors of the fingers flex and the extensor extends all those joints under their jurisdiction that are not secured by some other muscle or outside force. We have to bring the fingers and hand constantly into contact with external forces. This always happens when we are working on something, taking hold of something or carrying something, etc. This is why the form that is taken by the movement of the fingers depends more on the character of those external forces than on the muscles themselves taking part. The flexors and the extensor produce a raw force, crudely directed one way or another. The transformation of this force into a subtly elaborated working movement very often depends more on the implement and the method of taking hold of it than on the muscles of the hand themselves. From this you can already understand the enormous importance of taking hold of instruments correctly.

I am not going to describe the muscles of the thumb in detail. I shall only point out the main differences they have from the muscles of the other fingers. First of all, in the thumb, apart from the long muscles coming from the forearm there are also some fairly strong short ones; these are the ones that form the eminence of the thumb on the palm. The muscles of this eminence all somehow or other flex the thumb; that is why in the number of long muscles of the thumb there is only one flexor and all of two extensors. The second difference is that the metacarpal bone of the thumb (which is hidden in the soft part of the palm) can complete two kinds of movements in accordance with the two degrees of its mobility. In this it differs from the metacarpal bones of the other fingers, which are almost immobile. Consequently, it is served by its own staff of muscles; among them are one of the long extensors and some of the short flexors. All this gives the thumb significant and varied mobility.

With these facts we shall complete our survey of the muscles of the hand. It was necessarily very superficial; but I cannot wish for more than that you have, all the same, oriented yourselves in the definition and methods of action of the engines of the upper extremities. You can find a more detailed survey of the muscles of the hand and the form of their action in my book *General Biomechanics*. We will leave the lower extremity, which, by the way, is much simpler, to the next lecture as we have run out of time.

Lecture 6

Lecture 6 continues the detailed discussion of the working of muscles in relation to the lower limb and goes on to consider balance and the centres of gravity of the human body and its parts.

Comrades! We spent the whole of the last lecture looking at the muscles of the upper extremity. You probably are frightened by the thought of another rather boring lecture about the muscles of the leg, but we will go through the leg much faster, fortunately. These are the reasons why.

Firstly, you will remember that the upper or shoulder girdle is suspended entirely on muscles. There are rather a lot of these stabilizing muscles there and we have given them quite a lot of time and attention. Unlike the arms, the pelvic girdle is a hard structure and there is no muscle that works specially to secure it.

Secondly, our task of describing the leg muscles is simplified by the fact that leg and arm are very similar. We hinted a bit at this in the first lecture. Do you remember? In both, the top section is formed with one bone, the middle with two, and the most extreme one with a little "hand," which ends with five fingers or toes. The position of joints in the arm and leg is also similar: at the very top there is a three-axis ball joint; following it distally is a one-axis joint; and between the middle and the end segments there is a two-axis joint. Finally, there is another degree of mobility somewhere in the expanse of the middle segment (pronation and supination of the forearm, rotation of lower leg around its long axis inwards and outwards). Moreover, if you want to compare the forms of the corresponding bones of the arm and leg, you will see quite a similarity here as well. The similarity of all these features makes us ask the question: Is there a close similarity like this between the muscles of the upper and lower limbs?

If you have a look at an atlas of anatomy, alas, you will see hardly any likeness. You might think that the leg muscles are assembled according to a completely new plan, in which it is difficult to find any familiar features. The muscles of the hip and hip joint are especially bewildering. These muscles are a great torment for medical students, because there are a lot of them and they are placed at different angles to

© Springer Nature Switzerland AG 2020
N. A. Bernstein, *Biomechanics for Instructors*,
https://doi.org/10.1007/978-3-030-36163-1_7

Fig. 6.1 View of the right
shoulder blade from behind

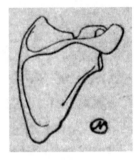

each other, they cross and intersect each other in every way possible, and in the final analysis it is not clear what they do. But do wait a little before you despair.

There is nothing surprising in the fact that you can't find unity in the plan for a human being. The purposes and the mechanical tasks that the upper and lower limbs have to carry out are too different. Even if there was once a general plan, then inevitably it was obscured long ago, and distorted because of the changing conditions. But let's look at a vertebrate whose two pairs of limbs carry out similar functions, to see whether a general plan is any clearer in them.

Firstly, let's look again at the bones and see which ones in the shoulder girdle correspond to which ones in the pelvic girdle. The shoulder blade, as you perhaps remember, consists of two parts: a plate that is turned towards the spine, that is, the back part, and two little rods turning towards the front part: the clavicle and coracoid process (Fig. 6.1). Each of the pelvic bones is formed in the same way: the plate goes towards the back and two little rods are turned towards the front. These rods are called the pubic and ischial parts in the pelvis. Both in the shoulder blade and in the hip, the articular hollow to which the upper bone of the limb is linked lies in the area between the little rods.

Also, don't forget something which was also mentioned in the first lecture. I mean the fact that the front and hind limbs face each other. You will see this clearly if you look at Fig. 6.2. Everything that looks backwards in the front limb looks forward in the back limb. Compare, for example, the elbow and the knee and the directions of the shoulder and hip bones in our diagram of the skeleton of a dog.

Let's begin by comparing the muscles of both limbs. The diagram that I just showed will calm you down a bit. Look at the great symmetry between the muscles of both limbs. In the diagram, the muscles that correspond to each other are drawn with the same sort of dotted lines, with the same numbers.

In the upper limb, we divided the muscles of the shoulder joint into two groups, short and long muscles. We can find similar groups near the hip joint. One of the long muscles of the shoulder (latissimus dorsi) starts far behind, from the spine, and stretches towards the head of the humerus, to be more exact, towards the lesser tubercle. The same type of muscle exists near the hip joint as well: it starts at the spine but far forward and stretches back towards the head of the femur (to be exact, to the prominence, called the lesser trochanter). In our diagram, this muscle is No. 1. The similarity of the two goes further, in that each has an addition, which starts from the plates of shoulder blade and of hip bone (No. 2). The muscle we are now describing is called iliopsoas and is the main flexor of the hip. And from the picture you can see its function more clearly.

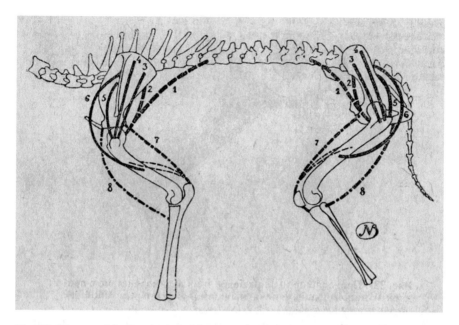

Fig. 6.2 Diagram of the organization of the muscles of the front and hind limbs of a four-legged mammal. The corresponding muscles are depicted by the same points and indicated by the same numbers

Fig. 6.3 Flexors (*C*) and extensors (*P*) of the hip joint and their action (by Mollier)

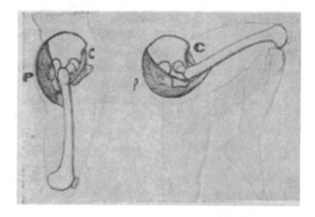

The same as in the shoulder region, this muscle has an opponent-antagonist. In the shoulder the antagonist is deltoid; here this muscle lies at the back of the hip, is huge in size (even more so than deltoid), and works as the extensor of the hip. It is called gluteus maximus (Fig. 6.2, No. 6). It is this muscle that forms the roundness of the bottom and can be seen very well outside on a human body. Figure 6.3 depicts both the iliac and the gluteus maximus of a human being and shows how they work.

There is also a muscle in the hip area corresponding to pectoralis major but it is very small and does not have any significance for us: so we won't pay it any attention.

Fig. 6.4 The muscles
which raise the thigh
outwards (*ПH*) and the
muscles "adducting" the
thigh (*ПP*) (by Mollier)

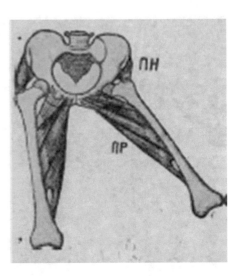

But the short muscles of the shoulder joint are reflected in the hip in a very pictur-
esque way and they have paramount importance.

Maybe you remember that from the low part of the shoulder blade plate, there
was a muscle that stretches to the greater tubercle, called infraspinatus. Anatomists
even distinguish here two whole muscles, which are closely linked (No. 3 and 4 in
Fig. 6.2). In the hip, the corresponding muscles (also No. 3 and 4) stretch in the
same order from the hip plate to the greater trochanter of the femur. Both these
muscles—the middle and small sciatic ones—raise the hip to the side. Figure 6.4
shows these muscles in a human being.

Finally, the small coracobrachialis muscle of the shoulder (from the coracoid
process of the shoulder blade to the inner surface of the humerus) has turned, in the
lower limb, into a huge packet of muscles, muscles which connect the pubic part of
the pelvis with the inside of the hip bone. This whole packet is called the adductors;
the name, of course, is wrong because in fact they don't adduct the thigh; they simply lower it from out inwards. These muscles are seen in Fig. 6.4.

We won't pay any attention to all the muscle trivia that are in abundance in the
pelvis; but it would be very easy to find corresponding muscles in the shoulder.
We'll do better to report on the four most important groups that we so far have found
around the hip joint:

1. Iliacus muscle—flexor of the thigh
2. Gluteus maximus—extensor
3. Middle and small glutei—raising the thigh outwards
4. Adductors—lowering the thigh inwards

We should note that the action of this mechanical group is in fact not as simple
and unified as it is described here. A lot depends on the initial position of the thigh,
the interaction of the various muscle groups, etc. It is possible to say that on the
whole, our four main groups can move the thigh in all directions to the limits of

its mobility. The small muscles that we have not described, while partly helping the main muscles, are at the same time responsible for the rotation of the thigh around its own long axis.

Let's also look at why the group of adductors is so large and strong in the pelvic girdle. The entire secret here is in those same properties of the arch that we have mentioned more than once. The weight of the body presses on the hip in the middle but the fulcra are on the sides. That means that the hip, together with the two legs, also forms an arch and as such it needs tension bracing below. Besides, we cannot pinion our legs together below of course; the tension brace is structured in a different way. From Fig. 6.4, one can see clearly that the adductors form this required tension brace. They don't let the legs slide apart under the action of the body; and their moment for this purpose must be quite considerable.

Let's move on to the muscles governing the knee joint. As in the elbow joint, there are only two groups of antagonists here. Don't forget that the muscle sitting on the back of the shoulder has to be reflected in the leg in the form of a muscle that sits in front and vice versa. So here, the three-headed extensor of the elbow corresponds exactly to the four-headed extensor of the knee (Fig. 6.2, No. 7). In the same way, the long head also goes through the ball (hip) joint. The short heads sit around the hip bone on all sides; only there are three in total here, not two.

There is yet another remarkable difference between the elbow and knee extensors. The elbow extensor ends below on the long hook of the elbow bone. With the knee extensor, it is as if such a hook got torn away from the tibia and got suspended in the middle of the muscle tendons right above the knee joint. This is a little bone that is called the kneecap. As you see, it is not an independent bone; it is like a bony callus that developed in the tendon where it goes around the joint and rubs against it.

It is interesting to add that the knee extensor is generally too short. It is enough for one joint, but it cannot always successfully serve both the joints through which it goes without them interfering with each other. Try to bend your leg at the knee and then extend it in the hip joint. When you get to the point where your hip can go no further, straighten your knee and you will see that the extension of the hip can be continued by 15 degrees more. You slackened the tension of your muscles at the knee and only then did it give you freedom of action in the hip joint.

At the opposite, the flexor side of the hip and knee lies in not two (as in the upper extremity) but three flexor muscles. One of them (indicated in Fig. 6.2, No. 8) corresponds exactly to biceps or the two-headed muscle of the upper arm and even has the same name: biceps or the two-headed muscle of the thigh. The first one attaches to the radius and this latter attaches to the fibula. One part of biceps femoris starts from the hip, and another part from the pelvis. Its action is obvious.

Instead of one internal shoulder muscle we have two in the hip. Both of them start from the ischial tuberosity, the same as the long head of biceps, and they diverge from it at the bottom in the form of a fork towards the inner edge of the tibia.

Biceps femoris and part of the four-headed extensor are shown in Fig. 6.5.

The flexors of the knee are also somewhat short. Here, this is even more noticeable than in the extensor. Let's do the same experiment, only the other way round. Straighten your knee and lift your leg up in front, that is, flex the upper leg.

Fig. 6.5 The two-headed
flexor of the knee (*A*) and
the extensor of the knee
(*B*) (by Mollier)

When you come to the limit and feel a characteristic pain in the hollow of the knee, then bend your knee. The pain will disappear immediately and you can flex your upper leg by 45 degrees or more. The reason is the same: insufficient ability to stretch in the muscles; as to why this is the case, try to explain it yourself.

In the lower leg, any similarity of the muscles with the muscles of the forearm is somewhat unclear even in four-legged animals. Therefore, we will abandon our scheme, which has helped us on many occasions, and we will try to describe the muscles of the knee briefly and clearly without comparisons.

One thing is remarkable as far as the lower leg is concerned. It is surrounded with very weak, thin muscles on several sides, which get tired easily and cannot carry out serious work. And suddenly, next to these feeble muscles a big, exceptionally strong muscle arises, like a mountain, which is responsible for extending the foot.[1]

So how can this be? Why is it that extending the foot has such inexplicable privileges? Why is there a giant muscle raised up next to dwarf muscles just here?

The reason is, in fact, that only this weight-bearing muscle carries out constant and responsible work. The rest are placed there as if just for form, because it is necessary to provide the foot with active mobility in those directions that are given to it by the mechanism of the ankle joints. But they rarely have to work and only then a little.

The giant that I just mentioned is the calf muscle—what we call the calves in everyday speech. This lies on the back of the lower leg and starts from the bones of

---

[1] Dorsiflexion.

Fig. 6.6 The calf muscle
and its role in standing (by
Mollier)

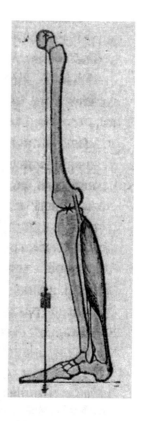

the lower leg and also from the lower ends of the femur. A huge lever serves as its lower attachment and this lever is formed by the heel bone, which projects far back. By the way, the lever arms of the other muscles of the lower leg are also as small as these muscles themselves.

Figure 6.6 explains to you the role that the calf muscle plays in standing. You see that in the majority of cases, the lower leg and sometimes also the thighs are inclined forward a bit. The center of gravity of the torso (we will talk about what this is and how to find it later today) turns out to be a little bit forward of both ankle joints. Therefore, the normal situation of things is that the whole body strives to fall forward. The calf muscles prevent this falling forward. During the fall forward we would flex the feet in the ankle joints,[2] and therefore the calf muscle acting as an extensor of the foot prevents the fall and balances the body in the standing position.

The role of the calf muscle is even more important in walking, and also running and jumping. Just observe yourself and you will see that all these movements are made with the help of a push off the ground by the toes, that is, with the help of extension of the foot. I should add that not only the beginning but also the end of a jump and the end of each step when walking and running largely depend on the

---

[2] Plantar flexion.

activity of the calf muscle. In these cases, it acts as suspension, softening and weakening all kinds of bumps. I advise you to ask the bus conductor what part of the body normally aches towards the evening. The suspension role demands from the muscles in charge a special constancy and tirelessness; and in consequence, the calf muscle is so strong and located in such a beneficial position. It is a wonderful worker and a person will certainly use it everywhere where only this is possible. Young women who press the bicycle pedals with the middle of the foot and not with the toes (and there are a lot of them in Moscow at the moment) behave like complete illiterates. In any type of work with pedals it is necessary to use the calf muscle. I should add that the pedal for the treadle sewing machine is also highly illiterate from the biomechanical point of view because its axis goes through the middle and it requires the use of not only the extensor but also the flexor of the foot and that is why, inevitably, the leg will get tired very quickly.

We will not have to consider separately the small muscles of the lower leg; they do not have great significance for us. I will only point out that some similarity to the forearm muscles is preserved here. As in the forearm, four muscles, which are in the corners, provide mobility to the two-axis joint between the lower leg and the foot. As in the forearm, the muscles of the toes are situated for the most part in the lower leg. Of course, the lower leg doesn't have any pronators and supinators. But, apart from this, the muscles of the toes receive much stronger support in the foot itself than the muscles of the fingers in the palm. The lower side of the foot is lined with a whole cushion of those muscles that play a significant role in maintaining the elasticity of the foot.

While some familiarity with the arm muscles was necessary for us to look at the arm as a working organ, our interest in the muscles of the leg is defined in a different way. While studying the working processes, the leg is interesting for us first of all as support for the body; as is sometimes said, what is important for us in the leg is not so much dynamics but statics. But in order for us to consider leg muscles and their purpose from this point of view, we have first of all to get acquainted with the main concepts of the balance of the human body and its parts.

Let's start with the study of the centers of gravity of the human body and its separate segments. As is known, the main property of the center of gravity of any body is as follows: it's a point through which the resultant of the force of gravity goes in every position possible for a given body. I'm afraid that it might be hard to understand this sort of definition. Let us approach it from another angle.

Let's take a bicycle wheel, which is sitting on its axis. Let's lift the bicycle up a bit and have a look at what is happening to the wheel. As you can see, it starts turning very slowly, completes a swing, then goes in the opposite direction, and then continues to make swinging motions back and forth which die down until at last it stops completely. What made the wheel start moving?

Students: Maybe there was some impetus at the beginning?
Lecturer: Maybe. Let's check. Here I am lifting the bicycle very carefully, holding the wheel with my hand. Now, already having lifted it, I carefully take my hand off the wheel: it starts swinging straightaway. Look at the wheel more carefully. It is not completely symmetrical: on one side

there is a metal valve through which we inflate the tire. I have with me another valve that has been broken off. I'm going to attach it to the wheel to the exactly opposite side of the first one. Now let's repeat the experiment; as you can see, the wheel does not even think about swinging.

So we have now a small number of observations; let's consider them from the mechanical point of view. The wheel of the bicycle, like any body, has weight, which means that it tries to fall downwards. This tendency to fall down is a property, of course, of each single element of the wheel. However, in our second experiment neither the wheel as a whole nor any element of it does drop down—consequently something is holding them back. Let us see: can all parts of the wheel go down at the same time? Obviously not, because we are holding the wheel at a certain height by its axis. That means, inevitably, that if some of the parts are dropping down then others are going up.

You know that work is measured by the product of the load multiplied by the change in height of this weight. The unit of work is a meter-kilogram, that is, work which is required for lifting 1 kg by 1 m.[3] If the same kilogram goes down by a meter, it will return the same portion of work. Let us turn our wheel to a certain angle. Some of its parts went up, that is, absorbed work, and others went down, that is, expended work. If the work of all parts that went up is equal exactly to the total work of all parts that went down, then the work of the wheel as a whole is zero. Because movement that is under the action of a force is always accompanied by real not zero work, then it means that no forces acted on the movement of our wheel. Therefore, the wheel is found to be in balance.

Thus, balance takes place when (and only when) the total work of the body with a small displacement, equals zero. In the case of a symmetrical wheel this rule is especially clear. Against each of these elements on the other side of the axis, the same distance from each other, there is another equal part, which always makes the exactly opposed movements to those made by the first part. The work of each such pair is always equal to zero when the axis is stationary.

In a wheel which has one valve, it is alone without a pair and consequently the force of gravity acts on it and makes it do work. It is necessary to apply a pair to it so that everything is in balance again.

What are the conditions then which are required for the pair of points to be in balance? We know that for this work that they carry when there is any displacement, the points must be constantly equal and opposite to each other. It's clear that the closer a point is to the axis of rotation, the smaller the displacements up and down that it can make. And because work is the product of the load multiplied by its vertical displacement, then in order to preserve the constancy of this product, one has to increase its first factor (the load) by the amount that you would decrease the second factor (that is, the extent to which the radius is lesser). We know that the product of

---

[3] Bernstein does not account for acceleration due to gravity in the definition of a force in the metric units he uses here. Normally this is discussed in terms of newtons, i.e. one newton is the force needed to accelerate 1 kg of mass at the rate of 1 m per second squared in the direction of the applied force.

the load multiplied by the radius is the moment of the load; that means that for balance, the moments of both points of our pair were required to be equal and opposite.

Let us turn to a case where it is necessary to balance out more than two points but three or more at once. Here the reasoning will be the same. You have already seen that when the moment of even a small group of particles that are concentrated in the valve differed from zero, that moment started immediately doing work and the wheel started moving. Therefore, we can draw the conclusion: in order for a body sitting on an axis to be in balance, it is necessary for the accumulated moment of all the parts of the body in relation to this axis to be equal to zero. The point of the body in relation to which the accumulated moment of all parts equals zero is the center of gravity of the body.

You understand that a body, which we are holding by the center of gravity and therefore keeping in balance, does not lose any part of its weight. If the wheel weighs 1 kg then all this weight of one 1 kg presses through the axis on the fulcrum. This means, first of all, that it is as if the action of gravity of all separate parts of the body can be substituted by the action of one single force, which equals the weight of our body and acts on the center of gravity of the body. Let's try now to support the body at some other point, to the side of the center of gravity. As before, the resultant of the weight of our body will act on the center of gravity but this time the point at which it acts will be aside from the fulcrum. Therefore, this time the force of gravity will have a certain moment. Take, for example, the weight of the body to be 1 kg and the distance from the center of gravity to the fulcrum to be 0.1 m (10 cm). The moment of the gravity in this case will be 0.1 m/kg and it is this moment that will define the movement of our object that is out of balance.

Let us now finish with our abstract discussions and attempt to apply our conclusions to biomechanics. Imagine that we have fixed our upper arm in space and the forearm can freely swing around the axis of the elbow joint. What will be the force of muscles moving the forearm in order to balance it or to move it? We can give an answer only when we know the weight of the forearm and the position of the center of gravity.

I won't tell you how scientists defined these weights and positions. In practice you should have an understanding of what these weights and their situations mean. However, in order to give you an idea about the ways of defining these, I will tell you about one of the old ways of working that were devoted to defining the position of the center of gravity of the body as a whole.

A person with normal build was put on a wooden board in a resting position. Then they put a bar, sharpened towards the top, on which the board could rock like a swing. Moving the bar forward and backward, they would find a position in which the board with the person lying on it would be exactly in balance. The position of the bar was drawn on the board after this. Then the bar was turned at a right angle and they looked for the position of balance again in the same way. The intersection of both found positions of the bar was obviously just under the total center of gravity of the human body and the board. Knowing the position of the center of gravity of the board one could easily calculate the position of the center of gravity of the body alone.

I will give you first of all an idea about the average values of the weight of the human body segments. These values are calculated for a person of normal build who weighs 60 kg. If you prefer *poody* and pounds calculate them yourself.

Names of the links	Weights of the links (kg)	Ratio of weights of the segments to the weight of the body
1. Head	4.236	0.0706
2. Trunk	25.620	0.4270
3. Upper leg	6.948	0.1158
4. Lower leg	3.162	0.0527
5. Foot	1074	0.0179
6. Upper arm	2.016	0.0336
7. Forearm	1.368	0.0228
8. Hand	0.504	0.0084
9. Head + trunk	29.856	0.4976
10. Lower leg + foot	4.236	0.0706
11. Whole leg	11.184	0.1864
12. Forearm + hand	1.872	0.0312
13. Whole arm	3.888	0.0648
14. Both legs	22.386	0.3728
15. Both arms	7.776	0.1296
16. Trunk + head + both arms	37.632	0.6272
17. Whole body	60.000	1.0000

Now we can consider how exactly the centers of gravity of all these segments are situated inside them. Here we encounter some difficulties. Firstly, the center of gravity of each segment, for example, the upper arm or forearm, never sits or stays exactly in one place. After all, muscles and skin are not immobile and they shift a little bit; consequently, the center of gravity shifts. Secondly, it's impossible to define in a living person where the center of gravity of the upper or lower leg is: and if we define these positions in some other indirect way (for example using a dead body), once and for all, who can confirm that these measurements will be the same as any living body? Fortunately, you don't need great accuracy and if we answer the question approximately then the small differences aren't very important. I will give you the following positions as rules that can be remembered easily.

If you draw in your mind straight lines along each of the long segments of the body, (that is, along the upper arm, forearm, and upper and lower leg), so that these lines go through the centers of both the most extreme joints of the given segment, then it turns out as follows. Firstly, in all these segments the centers of gravity lie on the lines that we have drawn, and we will call them the axes of the segments. Secondly, the distances of these centers of gravity from the centers of the extreme segments are in all cases in a ratio of approximately 4:5. The center of gravity is always a bit closer to the central joint than to the extreme joint.

As far as the hand and foot are concerned, everything with them works out fine as well. If we think of the axis of the foot as the line that goes from the end of the calcaneus towards the end of the second toe, then the center of gravity of the foot is on this axis and divides it into parts, in approximately the same ratio of 4:5. As far as the hand is concerned, due to its mobility, the center of gravity wanders about constantly and consequently we cannot say where its permanent place of residence is. Fortunately, the weight of the hand is very small (half a kilo) and consequently one can easily leave on one side all these small movements. If, as an approximation, we imagine the hand as fixed fast to the forearm, then one can be content with the knowledge of the position of the center of gravity of both those segments taken together. For this, we have to continue the axis of the forearm further to the tips of the fingers. Then the common center of gravity of the forearm and hand will be (approximately again, of course) on that axis and will be situated twice as close to the radiocarpal joint than to the elbow joint.

It is much more difficult to establish the position of the center of gravity of the torso: it is very flexible and constantly changes its form. We will mark the position of the center of gravity of the torso when in the quiet standing position. In this position the center of gravity will be found as follows. Draw horizontal straight lines to connect both shoulder and both hip joints. Divide the distance between the middle of these in a ratio of 4:5 (so that the top section corresponds to 4 and the low one to 5). Then the place of this division will be the center of gravity we are looking for. Finally, the center of gravity of the head is inside the skull, a bit higher than the line that connects both the ear holes.

All these data and methods may be fixed better in your mind, if you scrutinize Fig. 6.7 attentively. In this drawing the human body is depicted from the front. The whole area of the drawing is divided into small squares, each of which corresponds to 1 cm of natural size. The positions of all the centers of gravity that we just mentioned are shown in this drawing with circles.

I will have to return again to theoretical mechanics for a few minutes. We have to clarify how to find the center of gravity for a system that consists of two objects, if we know their weights and the position of the center of gravity in each of them. Surprisingly, this task is not as complicated as the task of searching for the center of gravity of each object taken separately. One should only remember that the center of gravity of a body is the point through which the resultant force of gravity goes. In other words, we can consider that the weight of each of our bodies is concentrated on this point. Therefore, the task which we have just set for ourselves turns into another one which is much simpler: the task of finding the center of gravity of a system which consists all in all of two points; moreover, the weight and the position of both points are known. And this problem has already been resolved in today's lecture. The heavier the load, the closer it should be to the center of gravity; the distance of the loads from the center of gravity should be inversely proportional to the loads themselves. Thus, the center of gravity of the system which consists of two weight points is situated on one of the straight lines with both points and divides the distance between them into sections which are inversely proportional to the weights which are concentrated on both points.

Look at Fig. 6.8. We can see schematically represented the elbow joint with the upper arm and the forearm with the hand. Their centers of gravity are indicated correspondingly with a_1 and a_2. The center of gravity of the whole arm must lie on the straight lines with both of these points, that is, at some point S. Its distance from a_1

Fig. 6.7 Diagram of a skeleton, showing the positions of the centers of gravity. *Γ*—center of gravity of the head. *T*—center of gravity of the torso. *Π*—center of gravity of the upper arm. *ΠR*—center of gravity of the forearm. *Б*—center of gravity of the upper leg. *Гол*—center of gravity of the lower leg. *ЦТ*—center of gravity of the whole body (by R. Fikks)

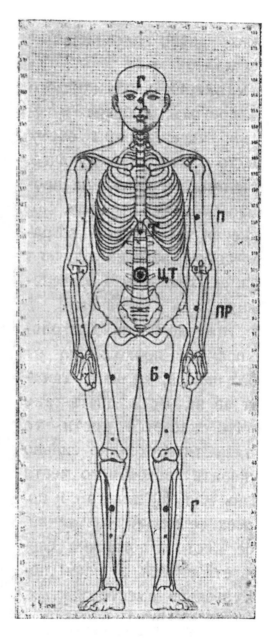

and a_2 should be inversely proportional to the weight of the upper arm and forearm. (In this example, we are going to consider the forearm and the hand together.) Thus the section r_1 is relative to section r_2, as the weight of forearm is relative to the weight of the shoulder. A very simple graphic technique for determining the position of such a common center of gravity flows from this.

From the center of gravity of the first segment a_1 mark in any direction the interval m_2, which shows in any units you like the weight of the *second* segment (Fig. 6.8). From

Fig. 6.8 Center of gravity
of two joined segments (*S*)
and the method of finding
it. The center of gravity of
A lies at a_1, and the center
of gravity of segment *B* at
a_2. The explanation of the
rest of the letters and lines
is in the text

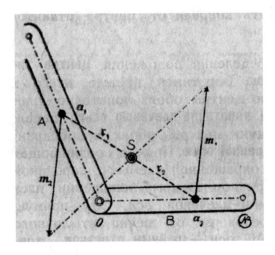

the center of gravity of the second segment draw a line which is parallel to the first line, but draw it in the opposite direction and on it mark the weight of the *first* segment, using the same units. Then connect, firstly, the centers of gravity of both segments, and secondly, the ends of the weight segments are just drawn with straight lines. The point of intersection of both these lines will be the common center of gravity *S*.

Clearly, this common center of gravity of the upper arm and the forearm does not stay in one place but changes its position depending on changes of elbow angle. Therefore, it is very difficult to keep pace with it. It is good that in your everyday practice you won't need that. The reason why I have told you about this method, nevertheless, is because it is needed today for another somewhat unexpected aim— that is, determining the optimal standing position. For now, remember that the center of gravity of our pair constantly divides the distance between the centers of gravity of each part in the same proportion.

This property of the center of gravity of a system gave the opportunity to a German scientist Fischer to construct a model with the help of which it is possible to determine for each position of the body and its parts the positions of the centers of gravity of each part and the whole body together. Such a model is presented in Fig. 6.9. You can see that the model is a hinged person who is equipped with a very complicated system of levers. The levers are arranged in such a way that places where they are joined consistently show the positions of the centers of gravity. Some of the joints are indicated in Fig. 6.9 with letters. *A* is the position of the center of gravity for the right arm, *B* the position of the center of gravity of both arms together, *C* the center of gravity of the trunk and head, *D* the center of gravity of right leg, and *E* the center of gravity of both legs together. Finally, *F* represents the position of center of gravity of the whole body.

We can see already from this model that the center of gravity for the whole body will change its position very significantly when the body is in motion. It would be extremely difficult without having at hand the model that is shown in Fig. 6.9 to know where the center of gravity is at every given minute. However, knowing the position and movement at least approximately could have huge practical significance. Therefore, I will teach you how to find these variables approximately for various instances.

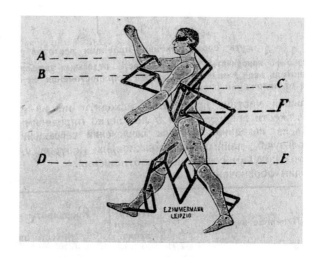

Fig. 6.9 Hinged person for determining the positions of the centers of gravity. Explanation in the text (by O. Fischer)

First of all, let's agree what's meant by the so-called normal standing position. This position is depicted in Fig. 6.7, which is ruled in squares, which we have seen already today. This corresponds more or less to the usual position at the front on the command "attention!," only without any tension and without any exaggerated sticking out of the chest. This standing position is all the more significant because in it, the centers of gravity in all parts of the body (apart from the feet) lie one above another in the same vertical plane.

In the normal standing position, the center of gravity of the whole body lies also in this central plane. You should remember that the centers of gravity of the segments are situated on straight lines that connect the centers of joints. With the normal standing position, both the centers of hip and knee joints and ankle joints are again in the same central plane.

The center of gravity of the whole body in a normal person in such a standing position is 4–5 cm higher than the line that connects the hip joints. If I show you how to find this line then at least you will know exactly where the center of gravity will be in a normal standing position. On the outer side of the hips, immediately under the edges of the pelvic bones, you can feel bony protrusions under the skin. These are, as a matter of fact, the greater trochanters of the hips, which we spoke about at the beginning of the lecture and which are shown in Fig. 6.4. If you find the superior parts of the greater trochanters on both sides and connect them mentally with a straight line, then this line will go through the centers of the hip joints.

Imagine now that a person has assumed the normal standing position and then changed the position of one arm without changing the position of any other part of the body. What will happen?

Look at Fig. 6.10. Two points of different weight *A* and *B* are depicted there and their common center of gravity is *S*. Let's imagine that *A* is ten times heavier than *B*; then the center of gravity *S* is ten times closer to *A* than to *B*. Now let's make *A* stationary and move *B* to new place *B'*. In our drawing, it is quite clear to you how the center of gravity *S* will shift. It will be transferred into the line which connects *B'* and *A* and will still be ten times closer to *A*. Let us call this new position *S'*. You

Fig. 6.10 Moving the center of gravity, S, of two points, A and B, of which one is ten times heavier than the other. Explanation in the text

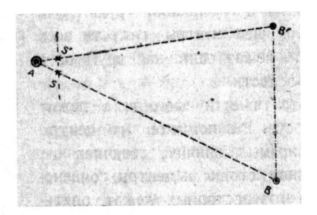

see that the triangles *ABB'* and *ASS'* are similar and moreover the sides of the first one are 11 times smaller than the second. Therefore, a shift of the center of gravity can be described in the following way.

If one part of the system stays immobile and another part moves in some direction by *n* cm then:

(1) The center of gravity of the system shifts in the direction parallel to the direction taken by the part that has been shifted.
(2) The magnitude of displacement of the center of gravity will be as many times smaller than *n* cm as the weight of the part that was moved is in proportion to the weight of the whole system.

Let's go back to the illustration of a person who has moved his hand. From the table that I already mentioned, we can see that the weight of the whole arm constitutes approximately 1/16th of the weight of the body. Consequently, if the person shifted the center of gravity of his arm by 16 cm in any direction, then the center of gravity of his body moved from the position which he had when standing normally, in the same direction, by 1 cm. Using the same reasoning, you will always be able to follow its movements, if the movements the person makes are simple.

From this rule, which in the final analysis isn't complicated and which we have arrived at by a roundabout route, another rule which is completely practical follows. This new rule indicates the method of defining the rational way of standing in a working operation.

I don't need to tell you how many arguments concentrate on the question of the efficient arrangement of the feet when working. You know very well the position required and proposed by TsIT and other similar establishments. You yourselves have several times asked me the question of what angle between the feet is more correct in sawing or chopping—67 or 70 degrees. Perhaps the extent to which all these values and proposals are random and unfounded is not so evident to you as it is to me, from my perspective. These values are often just made up in order to give some sort of firm rule and expound how scientific one's approach to the task is. However there is one method, very simple and scientifically uniquely correct, which, for some reason they do not use.

Fig. 6.11 The arrangement of the feet in normal standing. The circle is the projection of the center of gravity of the body. The area of support of the body is shaded in. The side of each square is 5 cm

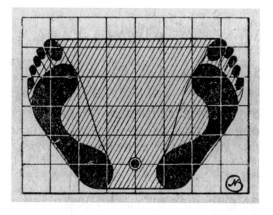

Let us begin again with normal standing. The average typical position of the feet in normal standing is shown in Fig. 6.11. This is in fact only the average position and the variation between one person and another can be quite considerable. Somehow or other each of us has a primary comfortable position for normal standing and we will start from this position.

In Fig. 6.11, a double circle indicates the position of the point where the plumb bob line falls downwards from the center of gravity. If the center of gravity is moved a few centimeters forward, to the right and so on, then the position of the drawn point will move on the floor plane for the same amount of centimeters and to the same side. We will call this point the projection of the center of gravity.

There is one basic rule about the center of gravity. This rule states: while the projection of the center of gravity is found inside the area of support, balance can be maintained; as soon as the projection goes beyond the boundaries of the area of support, balance is unconditionally destroyed and can only be restored if the area of support is moved afresh, so that the projection again is inside it. The area of support is so called, as you know, as it is the area included between all the extreme points of the body parts supported by the ground and by the straight lines linking these points together. You will get the area of support of the human body, if you join together by straight lines the ends of the toes and the ends of the heels of both feet. This area, composed of the area of support of each foot and of the space between them, is depicted by the shaded part of drawing 48.

It is clear that the closer the projection of the center of gravity of the body is to the edge of the area of support, the greater the risk of losing balance. Indeed, in such an extreme situation even a small deviation from the projection is enough for balance to be lost.[4] You see in Fig. 6.11 that in normal standing, the projection is generally located quite far back.

---

[4] Of course, even if the center of gravity moves outside the triangle of support, we don't have to lose balance completely but can regain it. If we are falling forwards we can bend our body forwards, or swing our arms forwards. This results in a temporary backwards force on the whole body after which we can then straighten up. The same in reverse is true if we overbalance backwards (think of the Keystone Cops swinging their arms!).

If the normal stance of the feet is adjusted well enough for quiet standing, that isn't the way it is with work. When you are working, parts of the body move and the center of gravity and its projection move immediately after them, and there is a danger that it will slip away from the territory of the area of support. What should be done in this situation? It is clear: you have to increase, to extend the area of support exactly in the direction where the center of gravity is wandering. And now you have to define not only that direction but also the swing of the movement of the projection.

Take, for example, chiselling. Observation of this movement shows that the general center of gravity of the hand and the hammer move in a plane which is at an angle of 40–50 degrees with the frontal plane. Their movement in this direction is 25–30 cm. That means that the center of gravity of the whole body will become displaced in this direction. The weight of the arm with the hammer is about 5 kg, i.e., approximately 1/12 of the weight of the whole body; consequently, the displaced center of gravity of the body when chopping is about 30/13 cm, i.e., about 2½ cm. If the movement of the hand with the hammer in chopping is slow and smooth, then it would be enough for us to extend the area of support exactly by 2½ cm and to extend it in exactly the same direction as the hand is moving in, that is, at an angle of 40–50 degrees towards the frontal plane.

In actuality, the movement of the hand in the chopping movement is quick and violent and therefore we need to increase the amount of displacement we've noted. I cannot put forward for you here all the necessary calculations; I shall only say that this quantity has to be approximately trebled, in order to balance also the moment of the momentum of the hammer. The effect of the swinging movement of the arm with the hammer in all its complexity is worked out as if the projection of the center of gravity of the body is moved by 7–8 cm.

A simple method for efficient standing when chopping flows from this. This is it. Stand in the normal, usual standing position (Fig. 6.11). Mark the position of the projection of the center of gravity of the body on the floor. From this position, draw a straight line to each side in the direction that the strike will go in. Then move each foot parallel with itself 3½–4 cm in the direction of this line—one in front, and the other behind. The stance arrived at in this way is indubitably better than what you are able to think up.

Obviously, all the arguments about 67 and 70 degrees completely fall away. It is possible to say boldly that in all operations where the ranges of the center of gravity are not very big, the angle between the feet remains the same as that which is customary for a given person in normal stance. Only the position of the feet changes. In operations demanding a larger range of the center of gravity (the swinging blow of the blacksmith's striker, the operation of a smoothing plane, etc.), the angle of the feet will change, but only because of the fact that when your feet are set widely apart it is biomechanically more useful to adopt a wide angle between the feet. In order to define efficient standing in a variety of working operations, we will investigate them in the form of seminars.

Lecture 7

Lecture 7 considers what animates the human machine, that is, the nervous system.

Comrades! In the preceding lectures we have investigated in sufficient detail the construction of the human machine and the rules by which these parts are assembled into a whole with the capacity to work. Now it is time to ask the question of what animates, what makes this artful mechanism move. We have investigated the mechanical parts of the ship named the human machine; now we must turn our attention to the captain's bridge. The central nervous system is this captain's bridge, that is, the sum total of the brain and spinal cord.

Before discussing the structure of the brain, let us consider what tasks it has to fulfil and what are the duties connected with the employment of the central nervous system.

First of all, there is an organ that is connected to all the smallest parts, organs and nooks and crannies of the human body. It has straight wires to almost every cell in the structure of the body. As we will see, this is a connection of a double kind, but the structure and the method of working of all these wires have the same basic features. These wires of the central nervous system are called the nerve fibres or nerves.

Figure 7.1 shows a view of a longitudinal section of a nerve fibre. What are called nerves in everyday speech are in fact whole cables of nerve fibres isolated from each other; each separate fibre is much finer even than those wires which are used for the winding of telephone electromagnets: they are one or two hundredths of a millimetre in section.

It is not accidental that I have compared the nerve fibres with isolating wire. The method of working and the construction itself are similar in many ways. Look at Fig. 7.1. You will see that the fibre has an axial rod clothed consecutively with two coverings. Physiological research has shown precisely that the conducting wire part, the wire in a specific sense, is only the central rod of the fibre, but the two coverings serve to do exactly what the winding of electrical wires does: both the one and the other are designated to separate, to isolate the fibres which lie next to each other from each other.

© Springer Nature Switzerland AG 2020
N. A. Bernstein, *Biomechanics for Instructors*,
https://doi.org/10.1007/978-3-030-36163-1_8

Fig. 7.1 Nerve fibres
under a microscope. The
conducting part is in the
middle surrounded by the
isolating cover (by
Rozenthal)

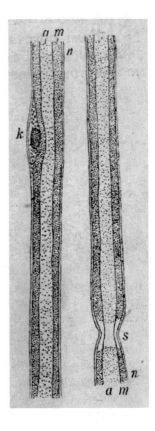

As you know, the telephone and telegraph wire connections take place with the help of an electric current that spreads along the wire. It is worth asking how it is that the connection between points of the human body through the connecting fibres takes place. Is the signal which runs along the nerve fibre similar to an electric current or not?

It is both similar and dissimilar. One thing is certain, that nervous phenomena are in essence actual electrical phenomena. The organism of animals and people has implemented an electrical connection in itself for hundreds of thousands of years before humanity hit upon such a connection in its states and cities. It is amusing to note that the human machine has also not got away from methods of intercommunication that are like the post, with its direct transport and parcel service; and the most amusing thing is that the 'post' in the human organism is regularly used by quite other organs and with other aims than those of the brain.[1] But the brain communicates with the subordinate organs of the body only by means of the nerve telegraph.

The difference between electrical processes in the wire and those in the nerve fibre is primarily the fact that in the wire, as we know, there is no movement of

---

[1] Bernstein's note: The glands of internal secretion.

particles of matter. In the wire along which the electrical current passes, the smallest particles of electricity—the electrons—are carried in one direction (from the cathode to the anode). I am sure that anyone who has been a radio 'ham' has an understanding of what electrons are. Here is such a signal carried along the nerve fibre and though it is also electric it is conveyed from place to place by particles of matter called ions. Each ion carries with it a charge, exactly equal to one or two (rarely more) electrons.

Those of you who study practical electrical engineering know, for example, that in inorganic nature there are also cases when the electrical charges are carried from place to place on particles of matter. Everyone who is engaged in electro-plating, nickel-plating, etcetera has seen that there are bubbles around the poles immersed in solution. If an iron object serves as one of the poles that are immersed in nickel salt solution, a fine layer of nickel is precipitated when the current is passed through. Where do the gas bubbles, nickel, etcetera come from? The theory of electricity shows that particles of nickel, charged with positive electricity, that is, positive nickel ions, are found in the solution and these ions are attracted by the negative pole immersed in the solution. Exactly these kinds of ions, only not of nickel but also of other lighter metals (mainly potassium and calcium), are always contained in the nerve fibre and are carried along its length with their charges. However, there is one more small difference between the electric current in the solution and the process that takes place in the nerve fibre. The fact is that when the current goes through the solution, in the end it carries great portions of matter from pole to pole and deposits this matter on the poles. This sort of depositing does not take place in the nerve fibre.

I will try to explain as simply as possible what takes place in the nerve fibres. I have never presented the most up-to-date views of this process to an audience such as you, who have comparatively little preparation for this, and so I am not sure whether what I say will be understandable enough.

You understand that research in the way it progresses in actuality cannot always begin with what is simple; it is connected inadvertently with what is easiest to discover or what happens most often. Consequently, the results achieved by research have to be arranged in order, and transformed little by little into a system and with difficulty. The view of the nature of electrical manifestations in the nerve that I wish to introduce to you has been worked out as a result of long and painstaking work by many scientists and represents the generalisation of many thousands of experiments.

It has been found that if in some point of the nerve fibre more positive ions have accumulated than in the neighbouring points, then the condition of this point changes in some way. First of all, as you will understand, this point then behaves as a positive pole; it will attract negative ions to itself (anions) and will reject positive ions (cations). This means that the electrical current flows near to this point from both sides.

Now imagine the resulting situation. If you can cope with this, everything will be clear to you. Let us call our positive point a momentary pole. It began to push the positive ions to the sides—this is because the positive particles are repelled by positive particles. But our momentary pole itself is positive only because there is a small surplus of positive particles in it. It dispels them to one side very quickly and remains with nothing; that is, it ceases to be a momentary pole. This means that the situation in the nerve is such that each disturbance in the disposition of its ions, that is, each appearance of the electric field, is immediately evened up by a small regrouping of ions, of which there is always a supply in the nerve. As you now have seen, this evening up takes place, because of the basic characteristic of similar electrical charges—they are repelled by each other.

What happens with the positive ions that are repelled from the momentary pole? Evidently, this. Next to the pole there were these ions (cations), exactly enough for balance, and now some other occupants have appeared; it is clear that the repelling begins here and now the cations begin to run away from this new point to the sides. In the same way, the process runs on further like a wave until it has been conveyed along the whole of the nerve fibre. So the electrical process in the nerve is the wave of oscillation of the cations around their middle positions, which runs along the whole nerve fibre. Scientists have even measured the speed of the course of such a wave. This speed turned out not to be very high—about 30 m/s, that is, about 100 km per hour. This is the same as that of a goods train at full speed.

This is the nature of the signal that the nervous system uses to link with its surrounding parts. I do not know whether it is worth talking about those physiological conclusions which result from what has been said; those of you who are interested in this further can very easily work out from this point of view all the numerous laws of activity of the nerve which have been established by science up to now and even foretell new laws. For example, from this theory it follows that the excited nerve fibre does not conduct the electrical current identically in both directions; that is, it can act as a rectifier of the current or a detector. Something in this direction has been discovered by the Russian scientist Verigo but we do not have time to spend on these interesting questions.[2]

The wave that runs along the nerve is called the neural excitation. Let us see how it acts on the muscle.

I already talked in the second lecture about how the muscle excitation takes place; now I will add that the muscle is excited each time the wave of the neural excitation reaches it along the nerve. You remember that the muscle contracts and works under the influence of excitation. The muscle in healthy conditions is never excited by itself; it has no right to take such a thing on itself. All muscle contractions happen obediently at the command of a higher power—the central nervous system.

A continuous conductor, called the motor nerve, extends from the spinal cord to each muscle, to each fibre. As the spinal cord can direct and organise the movement of all the muscles of the body from its place, you can imagine those nerve endings which are concentrated in the spinal cord as something like a keyboard, a system of buttons, which is necessary to be pressed in order to produce the movement of the muscles. We will soon see that the spinal cord is not the only one to undertake such

---

[2] Verigo, Bronislav Fortunatovich (1860–1925) was a Russian physiologist. He was a student of I. M. Sechenov and Georgian physiologist I. R. Tarkhanov (1846–1908) and later worked under the guidance of I. I. Mechnikov (1845–1916), zoologist and pioneer of immunology.

a complicated task and that the basic role in the directing of the orchestra of muscles is taken on itself by the brain.

All the time that we have been speaking of the nerve fibre it has been possible to regard ourselves as physicists. Regretfully we will soon have to move on from this sure and stable position. You will see now why this is.

Each nerve fibre of the motor nerve begins in the spinal cord from a head—a nerve cell. The nerve cell is fundamentally connected with its fibre and influences its way of acting significantly. Firstly, the very existence of the fibre depends on the wholeness of the cell. If the nerve fibre is severed two things happen: the part cut off from the cell quickly dies and the part remaining connected to the cell begins to grow, to penetrate into the isolating membrane of the cut and dying part, and often grows again to the former end of the nerve, that is, re-establishes its capacity to work. By the way, the method for stitching nerves together after injury is based on this; as long as the cells are whole the nerve can grow in exactly the same order that I have just described.

The second particularity of the fibre, which remains connected with the cell, is as follows. The nerve fibre can conduct the excitation in any direction by itself. The cell allows the excitation only in one specific direction through the nerve. The aggregate of the cell and its fibre is called the neuron; and so, thanks to the presence of the cell, the neurons are divided into two categories: some only conduct the stimulation from the spinal cord to the periphery only and others do only the opposite—from the surrounding parts of the body to the spinal cord. The neurons of the first type are called motor and the excitation they evoke is called centrifugal. The neurons of the second type are sensory and their excitation is centripetal.

The third particularity of the neuron in comparison with the separate fibre is that the fibre itself is as completely an indifferent conductor of excitation as the wire is the conductor of current. In contrast to this, the neuron is an independent source of excitation. It is a ready working unit (if you wish, a wire with a battery). The whole nervous system is composed of such tiny independent units; I think that it will be unnecessary to point out to you how small nerve cells are and what a large quantity of neurons comprise our nervous system. There are many milliards of independent neurons in it.

We have established the simplest form of the neural apparatus; now let us try to establish the simplest form of the nervous mechanism in a similar way. It turns out that the neurons also work in a very uniform way and it is not difficult to establish the basic type of their action. For now, let us bear in mind that the excitation of the motor neuron brings a specific muscle into action; the excitation of the sensory nerve appears when the sense organs (vision, hearing and touch in particular) perceive some sort of external impression and send a telegraph message about it to the centre.

As soon as an external impression has acted on the sense organs—light, sound, contact, etcetera—the corresponding signal is immediately directed to the brain along the nerve fibre. Following this signal there is, usually with great regularity, an answering signal to the motor nerve, which ends with a responsive movement of the animal. It is simplest to observe such answering movements in a low-order animal, for example, a frog. For such experiments a frog is decapitated. This is done for the following reason. I already said to you that the nerve cells of all neurons are connected with external organs—the sensory and the motor—that are concentrated in the spinal cord and around it. The brain is an additional complicated superstructure, which by itself does not stand in direct connection with one part of the body. Consequently, if we want to observe in a pure way the work of the spinal cord neuron, we must separate them from the interference of the brain. Mammals do not withstand decapitation, but a frog can live for a certain time without a head and so it is most convenient to work with it. It is possible to separate the larger part of the brain without harm to the life of a bird as well (for example, a dove) but this is technically more difficult and it does not take place so cleanly.

We hang the decapitated frog by the upper end of the body and then pinch its back leg. The leg immediately jerks. The same effect is achieved if you apply a burning match to the leg or dip the leg in acid, etcetera. For our purposes we do not need to describe what will happen in all similar circumstances with the frog; we would do better to make use of the simplest example, just described, in order to distinguish the mechanism which enables the twitching of the frog's leg to take place.

First of all, this is indubitably a mechanism and not a manifestation of will or something like that. The twitching comes with purely mechanical regularity and it would be too bold to speak about consciousness or will in a decapitated frog.

Secondly, the twitching takes place each time in response to the irritation of the frog's leg by one method or another; this is a responsive, reflected movement. Reflection in Latin is called reflex and consequently such reflected movements are called reflexes. And so in the experiment with the frog the brain can be excluded in an obvious way, so the twitching of the leg can be called a spinal cord reflex.

Human beings too have such mechanical, involuntary spinal cord reflexes. Who among you has not been at the doctor's when he/she has hit your knee below the patella with a little hammer in order to evoke an involuntary kick? Which of you has not noticed that the pupil of the eye narrows by itself when a bright light shines on it and that it widens in darkness? This movement is completely involuntary because we ourselves do not sense it and cannot voluntarily repeat it however much we might wish. Which of us can sneeze at will or hold back the movement of a cough for long? If any of you has had dealings with drunks, then you will know of one more form of reflex; a retching movement comes from the stimulation of the throat with a finger or a pen. A fully independent spinal cord is in charge of all this and many more spinal reflexes.

And so, with a reflex, excitation is conveyed as a kind of arc, which is consequently called the reflex arc. In the case, for example of the pupillary reflex, it begins in the eye with the action of light and is conveyed along the optic nerve to the upper part of the spinal cord (medulla), and from there it returns along a motor nerve into the eye to a small circular muscle which controls the movements of the pupil and brings about its narrowing. Figure 7.2 shows a diagram of the reflex arc of a simple spinal reflex. You can follow it along all its extent from the organs of feeling consecutively through two neurons to the muscle fibre.

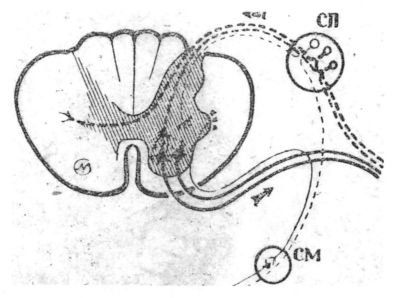

Fig. 7.2 The diagram of the reflex arc. On the left is a cross section of the spinal cord, on the right the nerve paths. The bold dotted line is the centripetal (sensory) neuron. The bold continuous line is the centrifugal (motor) neuron

How does the spinal cord, which we are now talking about, present itself? It is a braid, which is situated inside the bony canal formed by rings of the vertebrae. It stretches along this canal from the skull itself to the uppermost lumbar vertebra. The whole spinal cord presents itself as a huge, continuous accumulation of nerve cells, mostly belonging to motor neurons. If you cut a section across the spinal cord (Fig. 7.3), then the accumulation of nerve cells in this section will look like a brown blot in the middle, in the form of a butterfly. There is a shiny yellowish substance around the edges. Having looked at such a section under the microscope you will be sure that the yellowish substance is a mass of isolated nerve fibres, which our section has cut transversely. The section of these fibres bears a strong resemblance to a section of a telephone cable in appearance.

I shall tell you an interesting fact explaining the significance of the isolating membranes of neuron fibres. These membranes are composed of a fat-like substance, which in fact gives the nerve fibre a shiny white appearance. This substance, like all fats, is a poor conductor of electricity; the cables for currents of a million volts are not isolated by castor oil for nothing. Now you know that fats break down into spirit, chloroform, benzine and so on. In fact, how do you get a fatty mark out of clothing or a tablecloth? Evidently the membrane of a nerve fibre possesses the same qualities of solubility.

What happens if you drink a lot of spirits or ether, which comes immediately into contact with the membranes of the nerve fibres through the blood? To simplify what happens somewhat for clarity, these membranes begin to break down and their isolating properties weaken. What happens also is that when the isolation of any

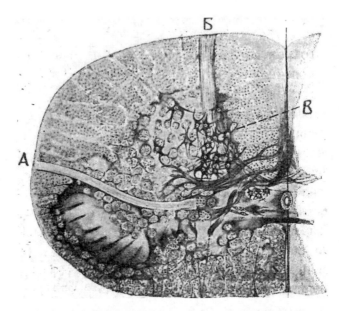

Fig. 7.3 Cross section of half a brain, somewhat schematised. (A) The outgoing bundle of motor nerve fibres, (Б) the incoming bundle of sensory nerves—(B), the nerve cells (by Deiters)

complex electrical wire is vitiated, a confusion of the currents results. The muscles will not move when and where they should, the messages from the sensory organs will be mixed up and will evoke unsuitable reflexes, etcetera. You look and you say, this person is drunk.[3]

Let us return to the spinal cord. Its brown substance (called sometimes the grey matter of the spinal cord) is an accumulation of cells, linked mainly with the motor nerves. Figure 7.3 shows how these motor nerves come from the ventral roots of the grey matter and each proceeds to their muscle. The sensory ones go to the dorsal side of the spinal cord and these cells are found not inside the spinal cord but around its sides.

The spinal cord is in charge of reflexes. Apart from simple reflexes it is not in charge of anything further. All the more complex movements and actions, which require the simultaneous and orderly participation of the many muscles, exceed the capability and possibility of the spinal cord. It acts with the exactness and monotony of an automatic machine: for everything that takes place beyond the boundaries of this monotony, it is compelled to turn higher, calling the brain into help.

A very large part of our movements and actions consists of simple reflexes, a much larger part than many of us think. Actually, simple reflexes are not only involuntary (that is, not only the half of them that is to do with movement flows below consciousness), but they are also unnoticed, since the sensory part of them

---

[3] Alcohol affects brain synapses directly, affecting balance, rather than dissolving away the myelin sheaths of nerves as Bernstein seems to be saying here.

also does not come into our consciousness. In the majority of cases, we do not pay attention to them and do not take them into account. Meanwhile, try to remember how many muscles are involved for example, in the simplest walking. We spent one and a half lectures on the enumeration of major muscles of the human body and indeed all these muscles somehow or other take part in the act of walking. How confused you would be if you had to consciously direct each of these muscles individually! Probably, you would not be able to take one step, and meanwhile you are not only not thinking about your legs at all when you are walking and what's more can be occupied with some completely external matter in this time: reading a book or making advances to a female companion.

Let us try now to systematise those branches of the control of movements that fall to the lot of the brain. It is usual at the beginning to describe the external view and structure of the brain and then to go on to its activity. With such a method of description, the brain seems extremely capricious and complex, it is confusing for everyone and it is impossible to understand any of it. So, I shall go by another path, constructively, and will begin with what I research, what, in fact, the brain has to do and how gradually and historically the necessary methods for the adaptation of this have developed in it. Those of you who do not mind looking at an external view of the brain all the same are recommended to look at Fig. 7.4 to find consolation in it.

It is evident that the central authority, no matter how complex the way it is organised, all the same will inevitably include in itself parts which receive the information from the periphery and parts which give the instructions to the periphery. Consequently, we must also in the first place seek sensory and motor apparatus in the brain.

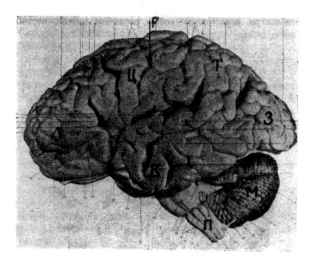

Fig. 7.4 View of the human brain from the left side. *Л*—the frontal part of the left hemisphere. *B*—the temporal part of the left hemisphere. *3*—the occipital part of the left hemisphere. *T*—the parietal part of the left hemisphere. *Ц*—the central part of the left hemisphere (where the motor centres are situated). *P*—the fissure of Rolando (now generally referred to as the central sulcus or central fissure). *M*—the cerebellum. *П*—the medulla (by Flatau)

You understand what kind of disorganisation would result if the supreme organ (equally the brain or the government, or the boss) decided to give orders directly to the periphery through the head of the nearest subordinates. This is avoided in the nervous system by the fact that not one nerve fibre from the brain goes further than the spinal cord and only the latter has direct wires to the muscles and organs of the body. The constitution of the human machine is very strict and not one petitioner from any part of the body has the possibility of directly relating to the upper power or generally to communicate, however that might be, otherwise than through its provincial government unit in the spinal cord. The white matter of the spinal cord that we revealed in its section is also a system of wires from the spinal cord towards the head and back. You remember that the sensory cells are connected with the posterior horns of the grey matter of the spinal cord and the motor cells lie in its anterior horns; consequently, it is evident that the ascending tracts of the spinal cord begin from the posterior horns of the grey matter and the descending tracts go down to the anterior horns.

In the human brain the sensory and the motor centres are to the right and the left of one main centre. It has already been said that the nervous system consists only of neurons, built according to one general plan, so clearly these centres are accumulations of cells, nerve endings of which go down and up through the white matter of the spinal cord. Those of you who have been scrutinising the external view of the brain all this time may be disappointed; neither those nor other centres are visible from the outside; they lie very deeply in the thickness of the brain and can only be seen having dissected the brain. For people who have a closer interest in the anatomy of the brain, I will say that the main movement centres of the brain are called the 'globus pallidus' and the main sensory centres the 'thalamus'. Both those and the others are visible in the section in Fig. 7.5. I will also add that each of these centres, both right and left, stand in connection with both sides of the brain—both the right and the left. Also, there are some even smaller centres of similar appointment, each with their own names, grouped around each of them but we will not linger on them and in our survey will think of them as joined up with the main centres.

Now think what would happen if these uniting centres acted completely independently from each other, not checking with each other and not agreeing their function with each other? Evidently, there, where the matter arises from a simple machine-like reflex to the interaction of hundreds of muscles, an independent and large centre is needed, destined specially for harmonising (as it is called, for co-ordinating) the particular nerve communications with each other. Such a centre must of course stand in connection both with the motor and the sensory centres of the spinal cord in order to have the possibility of regulating their action completely independently. Such a centre indeed exists in the human brain. This main harmonising or co-ordinating centre of the brain is the cerebellum.

So there is nothing up to now that is complicated or frightening. We have established that the link between the brain and the spinal cord conveniently fits into four nerve wires. This is the essence: there is the motor wire from the motor centre that goes below, the sensory wire goes above to the sensory centre and there are a pair

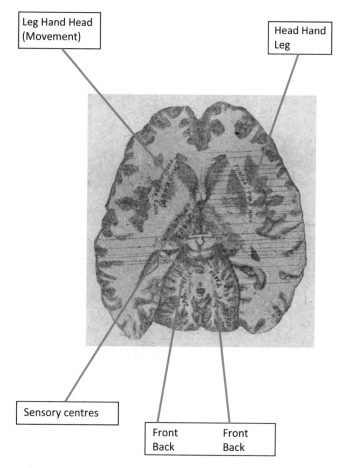

Fig. 7.5 A section of the brain with the front at the top and the back below (in the direction of a peaked cap band). Above there are the motor centres, in the middle the sensory centres and below the corresponding centres of the old brain (by Flatau)

of wires of both directions, uniting the spinal cord and the cerebellum. All these wires are also in actuality in the spinal cord; they are well illustrated schematically in Fig. 7.6.

The brain is just like this in the simply constructed lower vertebrates, for example, in frogs. It consists of two pairs of eminences behind which one unpaired thickening follows. The front two eminences are in essence the motor centres, the next two the sensory centres and finally the last thickening is the corresponding centre or the frog's cerebellum. In the human being, as we have already said, these centres, apart only from the cerebellum, are deeply hidden under the new vast brain formations, about which we will speak further.

This, in a few words, is how the reflex scheme becomes complicated after the cerebrum is included in it. Instead of a simple reflex arc, with only two links we

Fig. 7.6 A diagram of the conducting paths of the old brain. *3Б*—optic thalamus; *П*—globus pallidus; *М*—cerebellum. The triangles are the intermediate nuclei. The black triangle and the circle below depict the centres of the spinal cord with the centripetal and centrifugal nerves coming out from them

have a more complex one composed of at least four links—two centripetal and two centrifugal.

The general rule of work for the complex reflex arc is evidently this. It allows independently all those situations in which the spinal cord is in a condition to direct itself, but as we have seen, these situations are very few. They comprise, if you like, that background on which all the organised movements are carried out and which create the possibility for these movements. All the rest of these excitations, which, having arrived in the spinal cord, partly go back along the short arc, are directed from the centres of the cerebrum, and with their help responsive movements of a more or less complex type are stimulated.

There still remains a variety of opinions on the question of how the neurons of the cerebrum act on the spinal neurons. It is most likely that the actual motor stimuli originate in fact from the spinal cord and the cerebrum either moderates or inhibits these stimuli or on the other hand stops inhibiting them and gives them the possibility of manifesting at full strength. It is notable that in the decapitated frog the reflexes always appear more energetically than in a healthy one where the inhibiting part of the brain is not excluded. You know from your everyday experience that it is easier for the brain (and the consciousness linked with it) to inhibit some sort of reflex than to evoke it voluntarily. Everyone knows that it is possible to restrain oneself from sneezing, hiccups, etcetera, for some time, by the power of will, but no one can artificially hiccup or sneeze. (An actor's hiccupping is always imitation, only similar in the noise and the grimace.) There are illnesses where the motor link of the brain with the spinal cord is destroyed, as a consequence of which paralysis occurs. In the kind of paralysis where the link of a muscle with the spinal cord is preserved, that is, the spinal cord reflexes are not destroyed, the muscles are always in a tensed, stretched state and simple reflexes are even strengthened. If a nerve leading to the muscle is cut, so that the muscle loses the link with the spinal cord, then it becomes weak and the simple reflexes disappear altogether.

How the transmission of excitation from neuron to neuron takes place has not yet been established. Some consider that the finest fibrils, of which the core of the nerve fibre consists, are stretched along from one neuron to another through the nerve cell so that the reflex arc is an unbroken link from the organs of sense to the muscle itself. However, there are too many particularities of the nerve cell for us to agree unreservedly with this opinion. We have already seen that the nerve cell passes excitation only in one direction. Furthermore, it turns out that the excitation stimulus given by the nerve cell can be very strongly distinguished from the one that has come into this cell through a fibre of another neuron. It can appear stronger and weaker, it can appear interrupted and so on. As a whole, it turns out that the stimulus arriving in the nerve cell from another neuron is not transmitted to it even purely mechanically, but serves only as a signal, a summons, which brings into action the corresponding forces of the cell and makes it burst out with its own stimulus of excitation. Apart from this, only the spinal, simple reflexes are characterised by this mechanical regularity that allows the supposition of a strong link along the whole extent of the reflex arc. The reflexes of the cerebrum are characterised by their variety and adaptability. We will see further that many manifestations of activity of the brain can be explained only if we allow that switches exist in it and that its construction allows *temporary links* between separate neurons. We still do not know at all what these temporary links look like. Whether contiguity between the processes of the neurons takes place here or the stimulus is conveyed from one to another otherwise somehow—the resolution of this question is for the science of the future.

I have not said anything yet to you about the nature of those centripetal signals, which come into the brain from the sensory organs and evoke corresponding motor reflexes. Generally speaking, all the well-known sensory organs take part in the work of the reflexes; both that which we see and that which we hear and perceive can serve and do serve constantly as the stimuli for reflex movements. There is, however, one sensory system which in daily life people know little about; all the same it is this precise system that has the most influence on human movements. Those sense organs which are part of the structure of this system are called in physiology the proprioceptive organs; it will be more convenient for us to call them in Russian by a longer but more comprehensible name—organs of muscle joints and spatial sense.

These organs are distributed throughout the whole body. Most of them are in the tendons of the muscles and on the surface of joints. They can be in the muscles themselves. They are very small nerve endings that are visible only under a microscope.

A rather important task is allotted to the organs of spatial sense—to report to the brain about those movements and positions that each part of the human body takes.

Each of you can make any movement, write your signature, define the form of an object which is put in your hand, etcetera, quite accurately with your eyes closed (that is, without checking visually). All this is possible because of the army of sense organs that are distributed everywhere. These organs have their own nerve cable in the spinal cord which is directed to the thalamus. There is an illness called *tabes dorsalis*,[4] where this particular nerve path is destroyed; and immediately after its destruction, a serious disruption of movements appears. Such a disruption is always of the same order; neither force nor speed of movements suffer as a result of this disease but the regulation of movements and the ability to harmonise them are completely lost. In this case, a person loses stability, his/her gait is disturbed, his/her steps become excessively large and movements of the hands become jerky and uncertain. You can judge what is destroyed here by what these organs of spatial sense do in a healthy person.

In your instructor's everyday life the phrase 'muscle memory' is very useful. I make use of this occasion to point out to you a misunderstanding that is connected with this word. There is often talk about the visual or aural memory, meaning the sense connected with one or the other of the sense organs. There could also be talk about the muscle memory in this sense, as a memory connected with the activity of the muscular-articular sense. You could feel a stick with your eyes closed and then without opening your eyes measure the felt length on another stick. The memory that you display in this will be of course the memory of the muscular-articular sense, or muscle memory. There is something completely different intended by this word when people say that one or another working movement such as filing is mastered by a student through muscle memory. The mechanism, by means of which mastery of labour skills is gained, is much more complicated and is always based on the work of the whole of the brain. This is no longer muscle memory, but the general ability for motor learning and its focus do not lie in the organs of muscle sense nor in their centres but in the whole movement and co-ordinating centres.

In the system of joint muscle and spatial sensitivity, there is one apparatus, which is constructed in a particularly fine and complex way, which serves as if it is the main verifying observatory for the spatial sense. This apparatus is found in the skull, in close proximity to the organ of hearing on each side, and represents itself as two systems of subtle tubes, which go in various directions and are filled with fluid. This system of tubes acts as a kind of spirit level; only instead of a bubble of air, changes of pressure of the liquid on the walls of the tubes serve as a marker. This apparatus is the main organ of balance. A lot of interesting experiments with both animals and with human beings have been done on this, but regrettably we have too little time to mention them.

We have not yet finished our survey of the structure of the brain. I told you that in the form that has been described, the brain reveals full working capacity and adapts

---

[4] *Tabes dorsalis* is a degeneration (in fact, demyelination) of the neural tracts primarily in the dorsal columns of the spinal cord (the portion closest to the back of the body) and dorsal roots. These are the nerves that help maintain the sense of position (proprioception), vibration and discriminatory touch in a healthy person. Its manifestation is termed *locomotor ataxia*.

itself to life in some lower vertebrates such as for example, amphibians. In human beings it is complicated by one additional stage, which exists in a ready form only in mammals and complicates the system of the brain as we've described it up to now.

We have established that the brain of birds and amphibians consists of independent centres, motor, sensory and coordinating, and moreover each of the centres has its own link with the spinal cord. I now turn to describing the new formation, which has grown up in humans to such a degree that it has completely concealed itself and grown over all the parts of the brain described up to now. Everything that you see in Fig. 7.4 relates exclusively to this new brain, which is known as the cerebral hemispheres. I have no time today to talk about them; consequently I shall put off surveying them until the next lecture.

Lecture 8

In Lecture 8, Bernstein continues his survey of the nervous system with a discussion of the brain and then goes on to look at the study of human movement and the aim of biomechanics.

Comrades! In the previous lecture we didn't have time to finish the survey of the nervous system, short though it was, that I intended to give you. Even the little that you would need to know for your practice could easily take up a whole course specially dedicated to the nervous system. Since we are unable to give you such a course we have to bear with the fact that the information about the brain and its work that you will receive here will be very fragmented and incomplete.

I began to tell you about the cerebral hemispheres, which appeared only in mammals, in some sort of distinct form but then quickly developed subjugated to themselves all the remaining parts of the brain and acquired absolutely predominant significance in human beings.

The cerebral hemispheres are not constructed like the older centres that we talked about last time. In them there are no separate nuclei, distinct accumulations of nerve cells. The whole of the cerebral hemispheres is covered on the outside with a continuous layer of nerve cells of half a centimetre in thickness. This layer, enveloping the whole brain like a cortex, is therefore called the cerebral cortex. In a section of the brain the cerebral cortex looks like a chocolate brown layer, covering all the gyruses and sulci of the surface of the brain (Fig. 7.5). The expanse of this layer is very great: there are several times more nerve cells in it alone than in all the rest of the parts of the nervous system taken together. Figure 8.1 helps you understand how it looks under the microscope and how the nerve cells are distributed in the brain cortex.

The cerebral cortex so clearly predominates over all the other parts of the brain in the human being that for a long time it was the only thing about which something definite was known. It was studied earlier and in more detail than all the other parts of the brain and if I have now allowed myself to leave it until last then that is only thanks to the recent studies allowing us to establish the natural order of development and mutual subjugation of all parts of the nervous system. Therefore it will be more

© Springer Nature Switzerland AG 2020
N. A. Bernstein, *Biomechanics for Instructors*,
https://doi.org/10.1007/978-3-030-36163-1_9

Fig. 8.1 The cerebral
cortex under a microscope.
The black spots are the
nerve cells. On the right
are individual cells that
have been magnified (by
Campbell)[1]

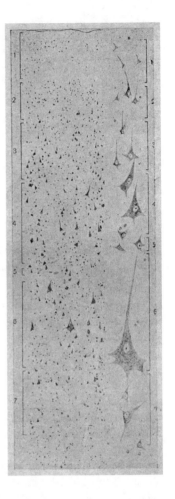

rational if I talk to you about the cerebral hemispheres not in the order in which it is studied and not in the way it is usually described but in the way most suitable for the scheme which we have stuck to, until now.

The cerebral cortex, just like the lower lying subordinate centres, concentrates in itself the motor and sensory sections. Here, true, they do not lie separately but are situated next to each other without sharp borders and perhaps even one overlaps another. At the same time, there is an interesting difference between the structure of sensory and motor parts of the cortex.

The sensory parts of the cortex are linked not only with the sensory organs (as was the case in the sensory centres of the old brain—the optic thalami) but also with the spinal cord. If it is possible to express oneself in this way, they are even further removed from life than the centres of the old brain. All that they have are wires from all the sensory sections of the old brain, that is, the optic thalami and sensory part of the cerebellum. These images of the lower sensory centres occupy a

---

[1] Alfred Walter Campbell (1868–1937) was an Australian neurologist who worked in London, Vienna and Prague.

Fig. 8.2 The conducting
paths of the new brain.
КП—cerebral cortex; the
other designations are as in
Fig. 8.6. *ПП*—pyramidal
path

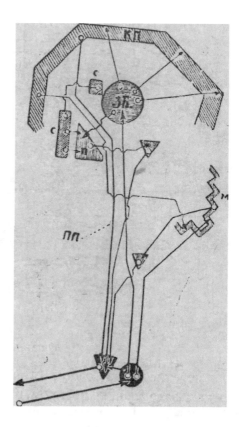

very major, dominating place in the cortex of the hemispheres. The sensory parts of
the cerebellum are represented primarily in the frontal parts of the cortex and the
optic thalami almost in the rest of it. This is shown schematically in Fig. 8.2.

The same image is shown even more clearly in Fig. 8.3. All the sensory centres
are inscribed there in italics. You see in this diagram the centre of vision (in the
occipital part) and the centre of hearing (in the temple part); as for the tactile and
pain centre, it is situated very widely in the central part at the back bank of a deep
gyrus called the fissure of Rolando. The names of parts of the body are in the diagram
where the corresponding sensory centres lie on the cortex.

As regards the motor centre of the cortex, it has behaved quite disobediently and
has demonstrated a particular lack of desire to be reckoned with and correspond
with whatever it could from the structure of the old brain. It is not linked in any way
either with the main motor centre of the old brain, the globus pallidus, or with the
motor sections of the cerebellum. On the contrary, it has to pave a fully independent
and specific path straight to the spinal cord, which communicates in such a way
directly with motor spinal cord cells. This path is also easily seen in Fig. 8.2. It is
called the pyramidal tract. In Fig. 8.3 the motor centre of the cortex is shown by

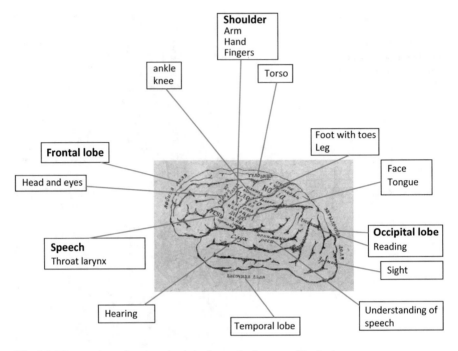

Fig. 8.3 The sensory and motor centres in the cerebral cortex of the brain

legends in printed letters. It lies on the front bank of Rolando's fissure, which we mentioned before, and occupies quite a big space. You can see that parts of this centre, related to the particular parts of the body, lie exactly side by side with corresponding parts of the tactile pain centres of the cortex. It is interesting that the centres of the upper parts of the body (the head, the tongue) lie in the brain lowest of all and the centres of the legs are situated at the very top.

There is another strange and until now unexplained characteristic of the cerebral hemispheres. It is this: all the centres of the left part of body lie in the right hemisphere and vice versa. All the conducting paths of the brain cortex, sooner or later, cross over from one side to another on the way. Thus the centres controlling the writing and the general activity of the right hand are situated in the left hemisphere of the brain. For reasons that are as yet unclear, the centres of speech are also in the left hemisphere so that the left hemisphere is in many ways more finely organised internally than the right. If destruction of brain matter (as a result of injury, bleeding, etcetera) takes place in the left hemisphere, then speech, understanding of speech, ability to write, reading, etcetera can easily be disturbed. When the right hemisphere and its nerve paths are damaged then paralysis of the left part of the body only usually takes place.

The pyramidal tract, that is, the motor path of the new brain, developed in animals later than all the other parts of the brain. This path also develops later than others in all children. Almost all the conducting paths of the brain cease their development already at the moment of birth, whereas the pyramidal path is ready for action only in the fifth or sixth month of a child's life. It is exactly at the point where the pyramidal path matures that the motor abilities of the child are enriched. The opening of new lines is marked by the fact that the child begins to learn to walk and to speak.

Earlier, people did not know about the activity of centres of the old brain; at best they were considered to be the intermediate points, which lay, for an unknown reason, on the paths of the brain hemispheres. Even less did they suspect how great the significance of these old centres is for the movements and the behaviour of a human being.

This significance is not fully evident and is insufficiently studied even now. At the same time, it is huge and it is impossible not to say something about it even very briefly.

You have already seen that the spinal cord reflexes comprise the most elementary, most general background of human movement. On the other hand, even from Fig. 8.3 you can get the impression that the activity of the cerebral hemispheres can be distinguished by its particular flexibility and many-sidedness and that it gives movement a final polishing. Experiments on the cortex of the cerebral hemispheres have shown that it has an extremely fine partitioning and adaptability. In dogs, it is easy to open the motor part of the brain cortex and to stimulate different points with an electric current. Each stimulation elicits movement of some part of the body or muscles of the dog (in exactly this way the first 'geographies' of the motor centres of dogs were obtained); and literally each new point of the cortex controls particular muscles and movements. There is nothing even remotely like such a subtle development existing in the motor centres of the old brain.

Thus, the task of the motor cortex consists, evidently, of the management of the most subtle and important part of the movements. The centres of the old brain give movement a coarse basic contour. You are familiar, of course, with moulding; it's possible to say that the old brain gives movements a cast and the hemispheres grind and polish this cast up to the margins of the required subtlety.

The second speciality of the cortex, which also results from its multi-partitioning, is the astonishing variety and flexibility of those movements, which it can evoke. It is exactly because of this that the ability to speak and write depends on the activity of the cortex; on it the even more varied and subtle work of thinking depends. On the contrary, movements, depending on the old brain, are uniform and monotonous; it seems that the majority of rhythmic movements come from the old brain. Walking, swimming, crawling and folkdance to a significant extent are directed by the activity of the old brain. Many uniform rhythmic labour movements depend on it to a large extent. The old brain has at its disposal a small number of movement formulae that are complex, but submit to change and re-education with difficulty. On the other hand, the 'library' of the new brain, as we have seen, is very rich, easily increases and is amenable to development. At the same time, the formulae, which the old brain possesses, present an extremely vast combination of movements, usually well adjusted and harmonised. The old brain works like an automat; it is not for nothing that many of these movements in which it plays the major role are called automatic movements.

What is called gracefulness, plasticity, etcetera, in the greatest degree, depends on the activity of the old brain; consequently, this is why it is not easy to train grace and it is almost always inborn. You will understand why people who are graceful and harmonious in large-scale movements do not necessarily turn out to be the most able to learn labour skills, for both one and the other are controlled by completely different parts of the brain.

With this we will finish our more than short survey of the nervous system. After the break, I will tell you how the study of human movement has been and is carried out, in which in fact the basic aim of biomechanics consists.

***

The aim of biomechanics is the study and correction of human movements. However, our course is so short and it is necessary to say so much that is preliminary that we can only come to the study of movements itself at the very end. All that has been explicated so far was only the prerequisites; this was the minimum store of knowledge needed; without this it is impossible to approach the study of movements. In essence, we have succeeded only in investigating the structure of the human machine; we have not yet observed it in action and up to now we do not know how to do this. Today, I would like to talk to you about what the science for the investigation of movements is and what methods are suitable for it.

Interest in movements appeared a long time ago, at the time, even when the only method of observation was simply by eye. Such interest arose earliest of all in artists who aimed to portray movement in drawing as accurately as possible. On the other hand, those inventors who wished to find the way to construct a flying machine (and there have been many of them through all times) assiduously studied the flight of birds, hoping to draw some instructions from this. In the fifteenth century there lived a person who was both a great artist and a great inventor. He was called Leonardo da Vinci. Many notes, drawings and measurements have been preserved in his note-books, concerned with questions of biomechanics, and in all fairness he is considered the first forefather of the science of movements.

However, the science of movements underwent great difficulty for a long time after Leonardo da Vinci because it did not have a precise method for the study of movements. Observations simply by eye are too untrustworthy, particularly when the concern is with quick and varied movements. Sketches are never insured against additions that are the fruit of the fantasy of the sketchers and it is impossible to catch them out in a lie and control them. Consequently, you will understand what a great boost the invention of photography was for the science of biomechanics.

In general, photography teaches people how to see. It is characteristic of human beings to be presumptuous; we imagine what we see much more than what is in fact there. Photography unmasked this proud self-deception. I suggest that you conduct an experiment like this: go into an art gallery and give your attention to the depiction of one object or another, for example, horses. Look at how horses were depicted up until the middle of the last century and how they started to draw them later. You will see an astonishing, enormous difference. After 1850 (approximately), you no longer see these eternal steeds on two back legs, playfully moving their front legs in the air. They begin gradually (though very slowly) to disappear and so does the

Fig. 8.4 Snapshot of a pole vault. Which of you readers has seen by eye such a position of the body? (from Lorentz photo catalogue)

depiction of horses galloping with front legs extended forward and back legs backward. What, indeed, can be said about the quick, fleeting movements of horses! Let us take pictures of a tree peacefully growing, or a house, mountains, etcetera. And you will also see here a difference between how they drew them before the invention of photography and how they started to draw afterwards. It will become clear to you how much there was that was a matter of convention in the old pictures. This autumn, in connection with the jubilee of the Academy of Sciences, there have been a lot of exhibitions in Moscow (the biggest was in The International Book)[2] where it was possible to see many old engravings depicting streets and houses. Do go and see whether nowadays any one is drawing in this false way or not. I do not want to assert that the new artists have learned from photography but the fact remains that with photography people understood for the first time how badly they saw. And once having understood, they made photography an integral part of each scientific investigation.

I will add further to what I have said that we not only see badly, that is, see little, but also see often what is not there and do not see what is there. You will all have had the chance to see snapshots of quick movements (Fig. 8.4). Has it never seemed to you that the position that has been photographed is completely unnatural and could never hap-

---

[2] In 1925, P.P. Shibanov, S.A. L'vov and P.N. Matynov researched and selected antique academic publications for a jubilee exhibition celebrating 200 years of the Academy of Sciences. The exhibition was called 'The International Book' in collaboration with the Academy of Sciences of the USSR.

pen in life? However, the photograph evidently cannot lie; this apparent unnaturalness is the best proof of how little it is possible in biomechanics to simply rely on the eye.

Why am I saying all this to you? The fact is that in your instructors' teaching practice you will constantly have to use visual observations and there will be hardly any times when there are more precise methods at your disposal. That is why I want to warn you a little, and give you a preventative inoculation so that you do not value your observational ability too highly. In order to demonstrate this, I will tell you a fact from my own biomechanical practice at TsIT.

There was an instructor there, who developed a new strike during chiselling, and assured everyone that this was the best one. And as unfortunately he was involved in teaching students, he taught them his method of chiselling. But in fact, this method, as precise biomechanical research has shown, has turned out not to be good but on the other hand very bad.

What's more this method was not only bad, but it was also impractical. Since the question of the suitability of this method was discussed very keenly at TsIT and the instructor was a hot-headed person, in order to convince him I photographed his own work many times. And then from all these photographs it turned out to be irrefutable that he was doing the chiselling with movements that were completely different from what he thought he was doing. It went so far as him being at the point of blaming the laboratory for the falsification of photographs, because his own intentions for movement and muscular joint impressions had departed from what the impartial photograph was showing him. We will later come back to the investigation of such photographs in the following lecture.

So do not allow yourselves to get conceited and photograph as often as you can what you are going to teach your students.

The invention of photography nearly 100 years ago has still not been much help in the matter of the study of movements. The first photographic plates had two shortcomings; they did not allow snapshots, which needed a lot of exposure, and apart from that could not be preserved very well, so that it was necessary to prepare them oneself immediately before the photograph and to put them in the camera still wet. It is understandable that in these conditions it was not possible to photograph movements but only artificially stiffened poses. I cannot refrain from giving one more warning here; when you wish to get acquainted with a movement that is unfamiliar (this particularly relates to those movements which are quick and sweeping), then do not make the person demonstrating them stop in the middle of the movements and show you a variety of poses in sequence. Almost all the poses will be completely different from those that the person adopts during a real movement.

It is only with the appearance of instant photography that the possibility arose of imprinting distinct moments of quick movements during their course. And then already, at the dawn of instant photography, one remarkable investigation into movements was conducted which can even now serve as a model.

Fig. 8.5 Photographs of a horse running, taken by Muybridge

Fig. 8.6 The fall of a cat, which had been held up by its legs, and how it turns while in flight (from Anschütz' photographs)

This research was conducted not by an academic but by an American horse trainer, Muybridge.[3] He was interested purely practically in the improvement of horse breeds and therefore gave himself the task of studying the gait of horses in order to have the opportunity of making comparisons between them.

His arrangement was cumbersome and complicated; Muybridge created a long shed. One of its walls was equipped with a counter that was open to the outdoors and on this counter about 20 identical photographic cameras were set up in a row. Each of them had an instant shutter, from which a long thread was extended. At some distance from the shed alongside it a fence of similar length was situated such that in between both a track was formed; and here the threads of the shutters were stretched across the track to the fence where they were fixed. All the cameras were loaded by photographic plates; after this a rider was seated on the horse and sent to gallop along the fence between the two dozen photographic eyes that were turned on him. The horse broke the threads with its chest one after another, and one after another the cameras clicked, at the moment as the horse ran past them. The result was a series of photographs producing the sequential positions of a horse running (Fig. 8.5). This entire huge and clumsy arrangement was, however, the first ancestor of contemporary cinema.

At approximately the same time as Muybridge, another researcher, Anschütz, was conducting photographic research of movements in Germany.[4] This researcher took many remarkable photographs of movements of animals and humans for the first time. Figure 8.6 shows one of his photographs. It depicts the movement of a cat falling feet upwards and turning in the air.

---

[3] Eadweard Muybridge (1830–1904, born Edward James Muggeridge) was an English photographer who pioneered photographic studies of motion, and work in motion-picture projection.

[4] Ottomar Anschütz (1846–1907) was a German inventor, photographer and chronophotographer.

At the end of the last century the photographic study of movements received a powerful impetus for development in connection with the work of the well-known French scientist Marey.[5] It is difficult to imagine anyone more inventive and resourceful. Marey studied the movements of all the animals he could, human gait, bird flight, etcetera, in his laboratory. For photographic studies, Marey often made use of a gun he had invented. This was a very strange gun; at the end of the barrel was a photographic lens and in place of the magazine, a drum, in which a round light-sensitive plate was placed. Marey took aim with this weapon at a running animal or bird and pulled the trigger and then ten sequential snapshots of the creature were made on the quickly rotating drum. There are two other methods of recording movements also attributed to this same Marey, which still remain significant.

One of them consists of this: the studied movements are transmitted to a springy drum, connected with a rubber pipe to a writing instrument. The very writing instrument is also a drum, metallic on one side and covered with a fine rubber membrane on the other. The pipe connects the internal cavities of both drums. As a result of the elasticity, the air pressure on the membrane of the first drum at the same time brings about the protrusion of the membrane onto the second drum. And in such a way, all movements, taken on the first, are transmitted to the second. With the membrane of the second drum, Marey connected a thin straw pointer, which was sharpened at the end, which could complete a rocking movement corresponding with each movement of the rubber membrane of the drum. A cylinder covered with sooted paper was placed near the sharp end of the pointer, revolving with a clockwork mechanism. When the sharp end touching the cylinder completed movements above and below, it traced its movement onto the sooted paper.

It would be difficult now to count the many cases of usage of adaptations of Marey's air transmission; it is used for the registration of heart contractions, muscle contractions, voice, etcetera.

Another of Marey's innovations had the aim of improving the photographic technique of recording movements. No one has contributed to the beginnings of kinematography to such an extent as Marey did, but in his time, the cinema was still only being born and could not be used for scientific work. Consequently, it was necessary to find roundabout routes.

Marey began to think about the question: Was it necessary, in studying human movements, to photograph it as a whole? Would it not be simpler to presuppose, as we have presupposed with you at the beginning of the course, that all the segments of the body are simple straightforward levers, without any internal mobility? And this presupposition put good methods in his hands.

You know that black objects do not work on photographic exposures; only light-coloured ones do. Marey dressed a subject in black velvet, with gloves and a hood, and made him stand against the background of a similarly black wall. The only light-coloured places on all the clothing of the subject were narrow silver laces sewn

---

[5] Etienne-Jules Marey (1830–1904) was a French scientist, photographer and chronophotographer.

Fig. 8.7 Photographic apparatus for cyclical snapshots and an electrical motor with rotating shutter (installation by the author at TsIT)

onto the external surface of his hands and feet. If you photograph a person like this, then from the whole figure, only those few light strips are shown on the exposure.

Marey made use of this, but he photographed on one and the same plate not once but several times in a row. He placed a cardboard circle, equipped with some openings (Fig. 8.7), in front of the lens of the photographic apparatus. If such a circle was rotated quickly, then it would at one point reveal and then at another cover the lens very quickly, if necessary, several tens of times per second. Snapshots would be taken as many times as that. Now imagine that the photographed person is not still, but, for example, is walking. Then all the consecutive positions that the sewn-on strips have occupied are photographed on the plate one after another, that is, the schematic positions of his extremities. Let our cardboard circle open the lines ten times per second; then there will be ten consecutive positions of the arms and legs each second on the snapshot. Figure 8.8 shows one of the photographs of human walking, taken by Marey.

Marey's method of simplifying the form of the object to some stripes in order to study his movements aroused general interest. Very soon in Germany two prominent scientists, Braune and Fischer, adopted a similar method for the most precise study of walking, to which they dedicated many years of work and six volumes of

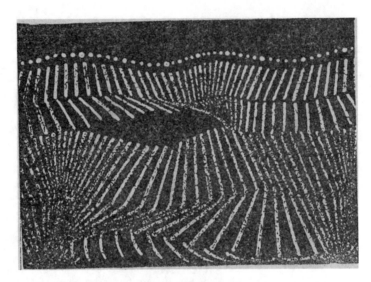

Fig. 8.8 Cyclogram of walking by Marey. The subject is walking from left to right. The movement of the head, right hand and right leg has been photographed (by O. Fischer)

essays.[6] Braune and Fischer used electric lamps instead of shiny strips in the form of slender tubes which made images on the plate in the form of thin lines and dots. The attire of the subject in their method was extremely heavy and clumsy; a photograph of walking by these scientists can be seen in Fig. 8.9.

Thus, two ways of studying movements photographically were outlined over 30 years ago. The first way consisted of obtaining as many separate instant photographs as possible of the whole of the moving subject. Muybridge and Anschütz worked on this method; Marey also used it in his gun; and from this the contemporary cinematograph arose. The other method led to the situation where nothing remained visible of the subject whose movement was being studied except a lot of nothing, that is, some lines and dots. Then the sequential positions of this simplified scheme were photographed many times in succession, in the same place. From this method, introduced as we have seen by Marey, the contemporary method of the cyclogram has developed, about which we will talk in the next lecture.

Now we will try to investigate what are the pros and cons of each method. By the way, at the same time I will tell you about the most recent innovations in each method.

The cinematic method has in its favour all the advantages of visual clarity. On individual cine photos, you have precise and detailed depictions of an object just as they appeared in actuality. If we pass a cinematographic snapshot through the

---

[6]Christian Wilhelm Braune (1831 Leipzig—1892) was a German anatomist and Otto Fischer (1861–1917) a physiologist. Inspired by Marey they conducted experimental studies of human gait and research on the centres of gravity in the human body as Bernstein describes.

Fig. 8.9 A cyclogram of walking by Braune and Fischer. The subject is walking from the left to the right. The point at the top is the crown of the head and the stripes from the top down are the upper arm, lower arm, hip, lower leg and foot; the square in the middle of the snapshot is the scale (by O. Fischer)

camera for demonstrations, you will again and again see that quick movement on the screen that has been once imprinted on the snapshot.

But all the advantages to the cinema of this visual clarity fall short in the case of the study of movements. For scientific work visual clarity is not a paramount advantage. Scientific research for the most part aims at measurability and exactness and in this regard, cinema, as we shall see, is now way behind the cyclographic method.

First of all, is cinema visually clear, as it seemed? Look at Fig. 8.10 where a series of cine snapshots of the strikes of a blacksmith's hammer are depicted. Try to say, from this drawing, what path the tip of the sledgehammer takes through the air. As you see, this is not so easy, more so because it is even more difficult to produce a precise measurement here. Figure 8.11 depicts the same movement but this has already been captured by the cyclographic method on one plate, and even, as you see, without anything sewn on or lights, only with the aid of the rotating shutter. Isn't it true that the path of the instrument on this photograph is indubitable and can be measured very easily and precisely? Further you will see that in many other respects the cyclo-photos turn out to be clearer than the cine photos.

The second advantage of cyclography over the cinema consists exactly in the fact that there are less fine points and details on the cyclogram than on the cine photo. If we wish to measure movement then we need to know exactly the movement of which point we are measuring. Consequently, all the details are superfluous and only obscure the essence of the matter; it is better to follow the movement exactly of three or four points than to get lost in the movement of several dozen.

Fig. 8.10 A cine series of the swing strike of a blacksmith's hammer. The upper line is the stroke and the recoil, the second and third lines are the swinging movement, the lower line is the striking movement (by Fremon)

Fig. 8.11 The same swing strike as in the previous picture but taken on one plate with the use of the rotating shutter. The path of the movement of the sledgehammer is very distinctly visible (by Fremon)

The third shortcoming of the cinema is in this. The cinema camera produces 16 photos a second; to take photographs faster than this is complicated by a number of technical considerations. But this is too few for quick movements, especially. In many of the cine photos that we took of the stroke, the contact of the hammer with the object that was being struck, that is, in essence, the most important moment of the whole stroke, did not appear on the plate. With the help of the cyclographic method, the quantity of the photos can easily be increased to 100 or more per second and at such speed it is already certain that all the most important details will appear on the photo.

Right at the beginning of the war, a new system of cinema appeared abroad, which permitted the production of not 16 photos a second but significantly more, about 400 photos a second. Photographic cameras of such a kind were given a very picturesque name, 'magnifying glasses of time'. And in fact, it is as if such cameras allow you to look at time through a magnifying glass. Imagine that we have photographed some sort of movement with a speed of 400 photos a second, then taken the same film with the photograph contained on it and passed it through a normal demonstration cine camera at a speed of 16 images per second. It will consequently show you the 400 photos which were done in one second, only in the course of 25 s in all; that is, it will show the movement 25 times more slowly than it was in fact carried out. In such a cine photo everything seems to be slowed down; a jumping horse sails smoothly through the air for a long time and in general all the quick, fine movements are carried out with unusual precision.

This fine camera would have had infinitely more advantages over any other method, if, along with its advantages, its shortcomings had not also increased. We must remember that a cine film is comparatively expensive and the magnifying glass of time devours it at four roubles per second. However, the ordinary cyclo-recording costs only a few kopeks for the same time. Also, the magnifying glass of time is a very big and complex device and very expensive, requiring the strongest artificial illumination, which makes using it in the everyday working situation impossible. Finally, the photograph taken by means of the magnifying glass of time is in the end a normal cine photo and is as inconvenient for measurement and processing as any cine photo. Perhaps it is because of this that as yet not one serious scientific study has been done with the magnifying glass of time.

Now I will tell you briefly how cyclographic photos are produced in a contemporary laboratory. It is necessary for me to give you an understanding of this in order that you might more easily understand the cyclograms that I will show you in the next lecture. The cyclogram method was developed for scientific aims by Dr Kekcheev[7] and myself; so I will explain to you those methods of work which I always use.

---

[7] Krikor Khachaturovich Kekcheev (1893–1948) was an eminent Russian scientist, an expert in physiology and psychophysiology of labour. From 1920 to 1925 he was the head of Psychology, the scientific secretary of the State Psychoneurological Institute in Moscow, then deputy director and head of the department of research (in the field of physiology of labour) of TsIT.

Fig. 8.12 A subject with small lamps marking his joints

We don't use either sewn-on strips or fragile tube lamps. We use those small electrical lamps that are used for pocket torches. We reasoned this through as follows.

If we're going to cut down an image, let's do it like this. Marey and Fischer did not need whole organs because their movements were sufficiently defined by movements of their long axes; but, you see, the movements of each straight line are defined as precisely by the movements of the two points situated at their ends. Therefore, we have no need to photograph the whole strip; we will limit ourselves to only its ends. Let us suppose that we know all the details of the movements of the centres of the elbow and shoulder joints; this is quite enough for us to establish immediately the movements of the whole upper arm. In fact, it is enough to join the position of the centres of both these joints with a straight line. It is true that by this method we will not catch the rotations of the upper arm around its long axis, but these rotations can always be defined if we know the movement of the long axis of the forearm. This means that we only need one point in the centre of the wrist joint.

There is no method for directly photographing the centre of the joint; consequently it is necessary to be satisfied with an approximate method. We place our small lamps on the body of the subject above the very centre of the joints, as near as possible to them. Figure 8.12 shows the position of the lamps for a photograph of the movement of the arm producing a chopping movement with a chisel. The lamps are placed above the shoulder, elbow and wrist joints, above the centre of gravity of the hand and on the centre of gravity of the hammer.

We dress a subject in a dark suit and place him against the background of a dark wall. We put a photographic camera with a rotating shutter in front of him. The photo itself is taken in the following way.

At first we illuminate the subject with strong lighting and take an ordinary photo of him in order to get a relatively clear image of his location and pose. After this we place only weak lighting in the room; we light up the lamps on the subject and put the revolving shutter into action. Then the subject is asked to produce the movements we want to photograph and when he has begun the photographic camera is opened and the little lamps imprint the trace of his movements on the plate.

We have to give ourselves also the opportunity to measure the recorded movement precisely. It is necessary to measure space and time as well. The following appliances are used for this.

At the beginning of the recording, we photograph a rod, divided into centimetres, with the person we are photographing. Since the distance from the camera to the rod and to the subject is measured beforehand as well, judging the natural size of the photographed movements is very easy.

We must also know how many images per second our recording will give. For this it is enough to know the speed of the rotation of the shutter. The speed of its rotation is measured by a very precise and simple method using sound and is also recorded.

Probably several of you have asked yourselves the question of whether the photograph that is giving two-dimensional images can reproduce natural movements that take place in space, that is, have depth as well as length and width. That's the case but for this there is the stereoscope. Here another defect of the cinema is revealed; today's cinema does not allow stereoscopic records. Besides, such a recording by means of the cyclographic method, using a normal photo camera, is quite possible. For this it is enough instead of a simple camera to place behind the shutter a stereoscopic camera, with two lenses and two plates. Then the movement will be captured in all its details.

In conclusion, I will show you one such stereoscopic cyclogram taken of chiselling (Fig. 8.13) and also the general view of the laboratory with an arrangement for cyclo-recording (Fig. 8.14). We will consider next time the methods for reading the cyclogram and in particular the analysis of the cyclogram.

Fig. 8.13 Half of the
stereoscopic cyclogram of
the chisel chopping
movement. The strike is
near to normal 8 (see
Lecture 10). Taken by the
author at TsIT

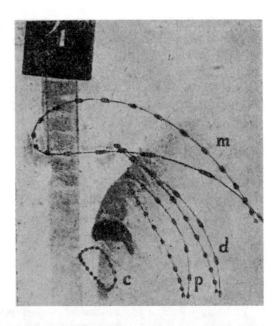

Fig. 8.14 The laboratory arrangement of the cyclographic recording (set up by the author at TsIT). *A*—Distributing electrical table where the direction of all the recordings takes place. *Б*—Photographic camera. *В*—Rotating shutter with four slots. *Г*—Light of 1000 candle power. *D*—Centimetre scale. *Е*—Workbench for the subject

Lecture 9

Lecture 9 continues the analysis of cyclograms that began in Lecture 8 in relation to walking and the hammer strike in chiselling.

Comrades! To begin, here is a cyclogram of walking for you (Fig. 9.1). Let's look today at reading the information contained in the cyclogram. At the same time, this will be a little exercise for you. The cyclographic method in its simplest forms is so uncomplicated and cheap that, as I hope, you will be interested in it purely practically. It is very possible that you will produce cyclographic recordings in industry yourselves; consequently it is desirable that you are literate in cyclographic terms, that you don't make the mistakes that often happen and that you have an understanding of what riches can be gained from a cyclogram, if you know how to handle it. The sin still lies on my soul that I have not yet written anything popular on the cyclographic method; consequently there exists the opinion that it is something very complicated which produces little. Of course, in any job there are complicated aspects and in scientific work one is allowed to give oneself whatever abstruse tasks one wishes. But in cyclograms, along with this abstruseness there is something of the ABC, accessible to all and which can come in useful at each step in the instructor's practice.

Let us consider our photograph. There are three dotted lines in all on it. These are the traces of the movement of three small lights. I attached these lights above the hip, knee and ankle joints of the right leg. The movement of the lights is a result of the fact that the person walked and carried the lights on himself. The direction of movement is from left to right. If you have taken in what has been said in the previous lecture, then explain to me why the traces of these lamps look like points.

Students: Because of the working of your shutter.
Lecturer: How did it work?
Students: It turns, then opens the lens and then closes it.
Lecturer: That's right; and the dotted line results because the light acts on the plate and leaves on it its trace only when the lens is open. Once it is shut then the light is not visible for the plate; besides, it continues to move

© Springer Nature Switzerland AG 2020
N. A. Bernstein, *Biomechanics for Instructors*,
https://doi.org/10.1007/978-3-030-36163-1_10

Fig. 9.1 A cyclogram of
walking photographed by
the author at TsIT. Only
the hip, knee and ankle
joints of the right leg have
been recorded. The
direction of walking is
from left to right. The walk
is a ceremonial march

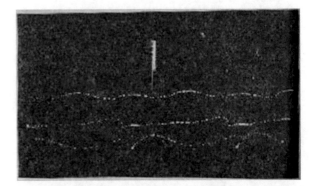

Fig. 9.2 A cyclogram of
chiselling, recorded
without a rotating shutter.
The vertical stroke (see
Lecture 10). Photographed
by the author at TsIT

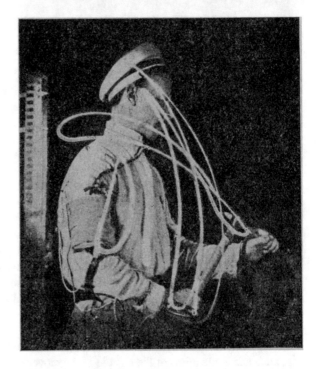

and when the lens is opened again, it then turns out to be far from that
place on which it was photographed last time. It is obvious that if the
shutter was taken away then there would be a continuous line on the
photo (Fig. 9.2). Now explain why the points are arranged in threes on
our cyclogram (Fig. 9.1)? You can't guess? Don't look for complicated
explanations; the matter is very simple. Imagine that our shutter is circular and three slots have been made in it situated as on a clock face
there are the numbers 9, 12 and 3. Such a shutter gives you an effect like
that achieved on our photo. If there were a fourth slot where number 6
is on the clock, then the points would follow one after another evenly
without gaps; and so from each of the four possible in fact one point is
let through. You will see later why this is so.

You understand that it is possible to arrange the slots on the disc as you wish; the appearance of the dotted line that results will change because of this.

Whatever the arrangement of points in the dotted line we have adopted, it is always possible to reconstruct the path or the trajectory of movements of each light easily, if you join all the dots that belong to its dotted line in a continuous curve. This is done, for example, in the cyclogram Fig. 10.4. Now ask yourselves the question: Why aren't all the dots from one of the dotted lines, let's say the upper one, the hip joint (Fig. 9.1), arranged at equal distances? Why in fact is one set of three closer together and the others are more spread out?

Audience:	The shutter has not rotated properly?
Lecturer:	No, the shutter has turned evenly; that's not it. Don't you think there might be a connection here with varied speed in walking? Perhaps in one case the hip light has moved faster, and in another more slowly?
Students:	Yes, of course.
Lecturer:	If you are in agreement with this, then what do you say: where will the movement be quicker and where slower?
Students:	It will be faster where the points are closer together (*arguments in the auditorium*).
Lecturer:	Is that so? Let us look at the photo of the hammer strike (Fig. 8.13). Here it is shown more clearly. The shutter has turned here evenly as always; its speed in the given photo was such that between each two points there was roughly 1/30th of a second. You see that in one place the points of the upper dotted line lie more closely together than in the other. There is a scale in centimetres on the photo. I take a ruler and measure the distance between the points according to this scale. In the widest place such a distance is 27 cm and in the closest 3 cm. Imagine this; in both cases the movement of points from one position to its neighbouring one continued for 1/30th of a second. Besides, in the first case it had in this time to go almost nine times further than in the second. Where is the speed of movement greater?
Students:	There, where it has passed more.
Lecturer:	Without a doubt and this is a general rule; the further the points were from one another, then the movement was faster. And can't you say by how many times the movement was faster in the first case than in the second approximately?
Students:	Nine times.
Lecturer:	Yes, that's approximately it. It's possible at the beginning to consider that the distances between the neighbouring points are proportional to the speeds of their movement. Now can't we calculate, although roughly, the speeds of the movement of the points themselves?
Students:	We can.
Lecturer:	How can you set about this?
Students:	You need to know the speed in one place. You need to compare.
Lecturer:	But it seems to me that you don't need to know anything; everything is already contained in the cyclogram. So, think, in 1/30th of a second the

point has passed 27 cm; that means that it has gone how far in a second? More or less?

Students: 30 times more.

Lecturer: How much is that?

Students: 810 cm.

Lecturer: This is how we've calculated the speed. In this place, it is about 8 m/s. Right now, let us formulate a general rule for the approximate calculation of speeds by means of cyclograms. In order to define the speed of movement in a given place, it is necessary to measure the distance between two neighbouring points, according to the scale on the photograph, and divide this quantity into the measure of time passing between two consecutive points. Of course, the scale must be located for this purpose at the same distance from the camera as the moving organ being filmed.

This method of defining speed is not very exact, though it suffices for practice. You can find a description of more exact methods in my essay, 'The investigation of the biomechanics of the stroke' in TsIT's 1923 *Collected Essays*. I think that you won't need these methods. I will add by the way that to define the speed of movement according to a cine film is very troublesome and inexact but the cyclogram gives it almost straightaway.

Here is a little practical conclusion for you from what has been said. The force of the hammer strike depends on its kinetic energy or living force. And kinetic energy is the product of the mass of the body and half the square of its speed. The mass of the body is equal to its weight in grams divided by 981.[1] Now if we know the weight of the hammer and its speed at the moment of the strike, then we can easily calculate the living force of the strike. For example, if the hammer (or more accurately its steel part) weighs 600 g and its speed before the strike is equal to 7 m/s, then the living force of the strike will be equal to 1½ m/kg. Consequently, knowing the weight of the hammer, you can define the power of the strike, that is, have an opinion about its productiveness, on the cyclogram.

One more thing is calculated very easily from this cyclogram. Think what happens when the shutter opens the lens? At this moment, evidently all the small lamps, however there are many of them, produce their image on the plate in the form of dots; if there were five of the lamps then there will be five dots and so on. At the next moment when the shutter opens again, all the dots are in new places and produce five new images. When, as a result of the recording, five varied dotted lines are achieved, it is then evident that each point on one of the dotted lines corresponds to a point on each of the remaining dotted lines. It is also understandable that the quantity of points on each of the dotted lines must be one and the same. If so, then let's find out the corresponding points (that is, the points that were photographed simultaneously) by means of some methods, on each of the dotted lines. If such

---

[1] Energy is given here as kilogram force times distance; this would now normally be expressed as Newtons times distance.

points on the two neighbouring dotted lines can be joined up by a straight line, then of course this line represents the position of the axis of the segment on the ends of which the two given lights sit. In such a way, we can sketch out also the position of all the rest of the segments we have photographed at that moment. What we get is no less than the diagram of the position of an arm or leg that has been photographed at a given moment. In the same way, we can establish from the cyclogram all the rest of the positions it takes consecutively. Here we have returned, from the cyclogram, which seems inexpressive at first glance, to the very things that the cinema has given us. Having in our hands such a series of sequential positions we can investigate a movement in a way that is no worse than if it was on a cine film. Figure 10.1 depicts the cyclogram of a strike developed in this way. For maximum clarity, its sequential positions have been slightly worked up in the form of a drawing. Its expressiveness leaves nothing to be desired. In Fig. 9.3 such a cyclogram has been developed in the form of a cine film. My colleagues and I prepared such a film and reproduced the movements of a chisel and the strike of a blacksmith's hammer in the form of stick diagrams.

And so we have already learned how to define from the cyclogram firstly, the path of movement of each part of the body, its dimensions and form; secondly, the speed of movement of each point; and thirdly, the sequential positions which a given organ occupied in space. In order to find out anything more from the cyclogram it is necessary for me to instruct you about one other thing in preparation.

You know from technical drawing lessons what is meant by a system of co-ordinates. We are using such a system now in our cyclogram. Let's have the hori-

Fig. 9.3 A section of cine film, showing chiselling in the form of an animation scheme. The film was prepared from the cyclograms

zontal directions signified by the letter X and the vertical ones by Y. We will divert ourselves from the third direction—the depth of the photo—for the time being, since it is defined only from stereoscopic photos and the development of such photos is in general much more complex and requires special apparatus. I refer those of you who are interested to the article I have already mentioned.

Let us take any two straight lines, vertical and horizontal, as the axis of the coordinates and we will measure the distance of each point of our cyclogram separately from one or another axis. For each point of light we will get two co-ordinates in this way: x-co-ordinate X and ordinate Y. Analysis of these co-ordinates gives us many new materials for acquaintance with the movement of the point.

Firstly, it could be interesting for us to find out something about the movements not of those points which have been captured on the plate, but of other intermediate points. If, for example, it is interesting to know what has happened to the middle of the humerus, then we can find out about this directly from the cyclogram. For this it is enough to join up with a straight line the corresponding positions of the shoulder and elbow lights and to divide the resulting line into two. But we often need other intermediate points; what do we do with them? It is very simple to deal with them. Let us suppose that the point we are interested in lies on one line between the two lights that have been photographed and divides the distance between them in a definite relationship, let's say $n{:}m$. Let the co-ordinates of one of the points that has been taken be equal to X and Y and the co-ordinates of the other be x and y. The co-ordinates of the intermediate point that we are looking for will be equal:

$$\frac{mX + nx}{m + n} \text{ and } \frac{mY + ny}{m + n}$$

As you see there is nothing intricate here.

A calculation of this kind could prove useful to us for example, in seeking the centre of gravity of the segments that lie exactly on the straight lines linking the centres of the joints. As you remember, in the long segments they divide these lines in a relationship of 4:5. Consequently, in all cases, if the coordinates of the upper joint (nearest to the torso) are in essence X and Y and the co-ordinates of the lower joint (furthest from the torso), defining the same segment, are equal to x and y, then the co-ordinates of the centre of gravity of this segment are in essence

$$\frac{4x + 5X}{9} \text{ and } \frac{4y + 5Y}{9}.$$

I will not expand here on how to find the centre of gravity of the entire multisegment systems by means of a cyclogram. Once we have found the position of the centres of gravity of individual segments, and once we know the corresponding weight of these segments (for this I have also given you a table of weights in one of the previous lectures), then to find the common centre of gravity for several segments won't be at all difficult. The general rule sounds like this:

If the co-ordinates of the centres of gravity of several segments in essence correspond to x_1, x_2, x_3 and so on and the masses (or weights) of the same segments are equal correspondingly to m_1, m_2, m_3 and so on, then the coordinates of the general centre of gravity of all the given segments will be

$$\frac{m_1 x_1 + m_2 + x_2 + m_3 x_3 + \ldots}{m_1 + m_2 + m_3 + \ldots}$$

If someone is interested more closely in these measurements, then he/she will find all the necessary numeric formulae ready to use in my book *General Biomechanics*.

I think that you understand yourselves the significance of precise information about the centres of gravity. First of all, by knowing the height to which the centre of gravity of the body is raised you can define the work expended in raising it. This definition is always necessary in the comparative evaluation of different types of working movement and we will return to it again when we study the correct and incorrect stroke. In the second place, familiarity with the centres of gravity helps us to clarify for ourselves the forces which are there in work and the calculation of it is a matter of great practical importance.

You probably know from general mechanics that force is measured as the product of mass and acceleration. We can now, using the table mentioned, define the masses of different parts of the human body. If we are able to define their acceleration by the cyclogram, then we will have in our hands all the materials for the calculation of forces. Let us see how we could define accelerations.

Acceleration is speed of change of velocity. In other words, acceleration is measured by the magnitude of changes in velocity per unit of time. I cannot discuss here the exact methods for calculating this by the cyclogram; but I will put into your hands a first coarse approximation. For this we will learn first to sketch the graphs of velocity of movement.

We have already approximately defined the approximate general velocity of a movement using the trajectory. Now I need to explain to you what are called the components of velocity. When a point moves, then both its co-ordinates, both X and Y, change. If we limit our consideration solely to changes in the x-co-ordinate (X), then from these changes it is possible to deduce what the component of velocity of movement was at any moment according to the x-co-ordinate. This is what this means. Let our point move diagonally with a speed of 5 m/s. However, it could turn out for example that the x-co-ordinate changed at a velocity of 3 m/s and the ordinate at a velocity of 4 m/s. If both these components of velocity add up according to the rule of the parallelogram of velocities then, as a result, exactly the original velocity of movement along a trajectory, that is, 5 m/s, will be achieved. In consequence, not going further than that approximation which we used already in the calculation of common velocity, we can say the following. If the shutter gives us an image with 50 points per second and for one 1/50th of a second, the x-co-ordinate of the point has changed by n centimetres, then the component of velocity of the point on the x-coordinate consists of $50\,n$ cm^2. We agree to consider

---

[2] This should be cm/s.

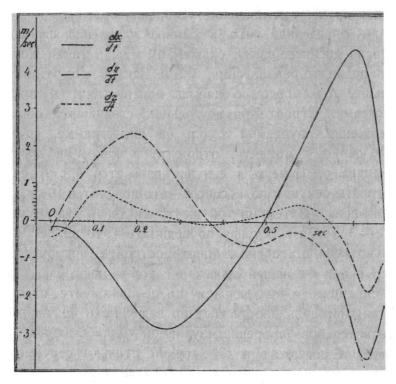

Fig. 9.4 Graph of the component velocities of speeds of movement of the hammer in chiselling. _____ velocity of movement forward and back; - - - velocity of movement up and down; .... velocity of movement right and left

that if the x-co-ordinate is increasing then the speed is positive; if it decreases the speed is negative. The same consideration also relates of course to the component velocity on the y-coordinate.

Now we can come to the composition of the graph of components of the velocities. Plot two mutually perpendicular axes of the co-ordinates, along the y-axis put the value of the speeds (for example, metres per second), and along the axis of the x-coordinate indicate a fraction of a second. Such axes are in Fig. 9.4, for example. Then, for each moment of time put beside the axis of the x-co-ordinate, perpendicularly to it, the values of the velocities, which are found at the corresponding moment; put positive velocities at the top, and negative ones below. If you unite the ends of all the sections that have been put aside with a curved line, then you will get the same graph of velocities which we wanted to achieve. Such a graph of velocities of the centre of gravity of a chiselling hammer is shown in Fig. 9.4. The continuous curve is a component velocity of movement forward and back; a dashed curve is the same for the component velocity in the up and down direction; at the end, there is a dotted line—the same for the movement right and left.

What do all these graphs do for us? What do they give us that is new? As you will soon see, there is very much that is new. Through them we will move to a definition of forces by a straight path.

We have already noted that acceleration is the rate of change of velocity. We already have the velocity, in a calculated and diagrammatic form, according to all the coordinates. Now we can reason thus. At a certain moment, the velocity of the point on the x-co-ordinate comprises 2 m/s; after, let's say, 0.1 s, it, as it emerges from the graph, has achieved already 2½ m/s. In 0.1 s it has changed by ½ a metre; consequently at this tempo of change, that is, with this acceleration, it would be a change in a whole second of a whole 5 m. Consequently the acceleration at the given interval comprises, in a rough approximation, 5 m/s.[3] It is possible to calculate the acceleration for each interval generally, just as exactly. Only do not mix up the signs of positive and negative accelerations. Acceleration is positive, if on the graph of velocity the curve goes to the right and up; it is negative if the curve goes from the right downwards. You will soon get used to the graph of accelerations and you'll notice that the steeper the curve of the velocities, the greater the value of the acceleration.

Here you have defined that at a given moment the component acceleration on the x-co-ordinate is equal, let's say, to 10m/s and the same component according to the ordinates is equal, for example, to 3 m/s. What more can be done with these numbers? Again, we need to unite them according to the rules of the parallelogram. Plot a rectangle in this way: from the depiction of the point in question on the cyclogram that corresponds to a certain moment, separate off into some scale (for example, 1 m/s is equivalent to 1 cm), firstly + 10 m/s on the x-co-ordinate—this will result in 10 cm horizontally on the right. Then from the same point and in the same scale set aside—3 m/s on the ordinate—this will result in 3 cm vertically down. Construct a rectangle on both lines, take the diagonal into it from our beginning point, draw it more thickly and put an arrow at the far end. The resulting diagonal will depict in size and direction the actual acceleration of the point being studied at a given moment; the arrow contains approximately as many centimetres as the full acceleration of the point contains in a second. It is now not difficult to get from the acceleration to the force. After all, the direction of the force coincides with the direction of the acceleration and the value results if the value of the acceleration is multiplied by the mass of the point. This means that the force can also be depicted as an arrow, having chosen a suitable scale for it.

All the calculations that have been described might seem to you to be laborious and boring. On the other hand, how fascinatingly interesting it is when out of a lifeless cyclogram, which is like a motley net of points, suddenly one by one all the secrets of the movement which has taken place begin to appear! There is no cine film that will give you even anything like this wealth of information about movement that the cyclogram has been able to give you, after it has been skilfully worked out. You begin to feel that you have learned to read some language which

---

[3] Acceleration units are in fact m/s². This applies here and in the next paragraph.

was incomprehensible to you before; and the pleasure of reading this language page by page is so great that all the unpleasantness and all the boredom of the preparatory calculations are left behind.

I will show you a picture in which I have calculated and drawn the forces in the centres of gravity on the hammer strike according to a cyclogram. I will not print this drawing as it has been printed so many times that probably everyone is bored with it. You will find it in Dr. Kekcheev's book, *Fizilogiya Truda* (*The Physiology of Labour*).[4]

Now we are equipped with some cyclographic grammar and can come to investigate a labour movement in all its details. But today we have little time and so I will leave this investigation until the next lecture and now I will acquaint you briefly with the varied fields of application of the cyclogram method in order to convince you how widely and variedly they may be used. At the same time, it will do no harm to have a rest since the material I have given you in today's lecture was inadvertently rather exhausting.

It is difficult to imagine any field of labour movements where the cyclographic method cannot be used with success. In my collection, there are cyclograms of the most varied operations.

However, it is not convenient to photograph all working movements on a simple cyclogram. Take filing, for example. In this work, the movement is performed back and forth in one direction, in one and the same place. If you try to photograph such a movement on a cyclogram, then all the points will lie on top of each other and it will be absolutely impossible to make anything out. In order to save the situation we should resort to an original trick.

In order to photograph this kind of movement, we constructed a special camera at TsIT (with the help of engineer A. Yalovyi).[5] In this the plate was no longer immobilised but was able to move with a uniform movement along the lens. If the snapshot of the forward and return course of the file fell in the same place, then this sliding plate has time in between the forward and return course of the saw to move a little distance. As a result, instead of a smudged line an extended clear line is achieved. Figure 9.5 depicts one of the cyclograms of filing photographed in this way.

This method with the sliding plate is extremely rich with possibilities for another reason. On the normal cyclogram, it is impossible to photograph more than one of a series of repeated movements at once. In the opposite case, the photographs of the points of the second movement come together in the same places as in the first movement and an illegible daub results. However, if such a rhythmic movement is photographed on a sliding plate, then it is possible to achieve as many cycles of movement as possible one after another. In our picture of filing, more than four cycles are accommodated; and if the plate is changed for a film rotating around a

---

[4] K. Kh. Kekcheev, *Fiziologiya Truda*. 1925.

[5] A. A. Yalovyi, part of the staff of TsIT, was the author of 'Metody svetovoi zapisi raboty pri rubke zubilom' (Methods of light registration in chiselling). *Issledovaniya TsITa*,1,2,1924 [in Russian].

Fig. 9.5 A cyclogram of filing, taken using a camera with a sliding cassette. From left to right the traces are (1) the right elbow, (2) the right wrist, (3) the right index finger and (4) the left thumb. Taken by the author at TsIT

bobbin then the quantity of cycles can be increased to any limit. I describe such a film camera in the second collection, *Voposy psykhofiziologii, refleksologii i gigieny truda* (*Questions of psychophysiology, reflexology and the hygiene of labour*).

It is possible to see in the photograph under discussion how strikingly similar the consecutive movements of an experienced worker are. I made particularly subtle measurements of this kind of graph and was convinced that the difference in duration of separate cycles in one sawing does not usually exceed more than two hundredth parts of a second.

The same camera with a sliding cassette can be used in other situations. The use of the cyclograph for clinical study is not the subject of our course. That is why I cannot allow myself to give you a picture; but with the help of this method I have received very interesting notes of movement of people who are ill which provide much for discussion about their illness and for diagnosis.

Finally, I shall tell you about one more use of cyclograms. In defining professional aptitude we often meet with the question of how well developed in a person is the sense of space which we discussed in one of the last lectures. The availability of this sense in the greatest degree possible is needed by pilots, chauffeurs, miners, people to whom it falls to measure distance or to work in darkness and so on. In the profession of metal workers there are also many cases when a well-developed spatial sense can do a good turn and even prevent a person from misfortune.

It is handy to research the sense of space, for example, in this way. A circle or a triangle of 3–4 m in width is drawn on the floor. A subject is asked to go along this figure in order to learn it precisely. Immediately after this a bandage is tied over his eyes and he is asked to walk in the same figure blindfolded. The better developed the sense of space in a person and the less labile (unstable, subject to disturbing influences) it is, the more correctly the person will fulfil the task and the more insignificant will be his deviation from the drawn path when walking with closed eyes. Here the question is only of how to register the behaviour of the person being tested, and how to study the deviations which result.

Right here cyclography comes to our help again. The experiment is set up as depicted in Fig. 9.6. High above the floor a photographic apparatus is set up with the lens directed straight down. Only one lamp is firmly fixed on the top of the head of the person being tested. After this, the figure drawn on the floor is photographed from a height, the person's eyes are bandaged and he sets off while the camera from above follows the movement of the lamp on the top of his head and photographs its path. In Fig. 9.7a, b, there are two photos taken from a height in the way that has

Fig. 9.6 Diagram of a set-up for photographing adapted by the author for experiments with the spatial sense at TsIT. *A*—Photo camera, *Б*—A subject, *B*—observer

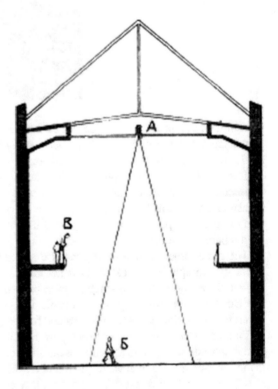

been described. In both cases the task was one and the same: a triangle. Meanwhile you will see what a big difference there is in the execution of the two people being tested; the first was a completely healthy and normal person with a well-developed spatial sense; the second had had typhus not long before which had affected his nervous system. Such a test of spatial sense enables the selection of people suitable for a given profession in an extremely vivid and precise way. And these experiments can of course be varied infinitely, but I think that what has been said is enough to give you an understanding of how wide the field of application of cyclography is in the study of movements.

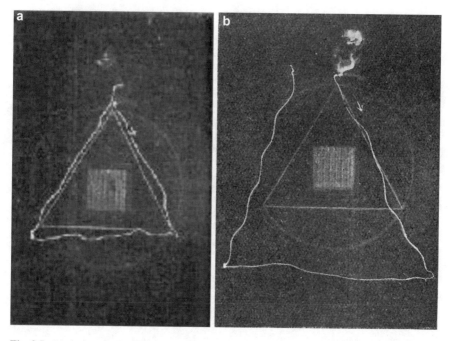

Fig. 9.7 (a) A photograph of walking blindfolded in a triangle, taken from a height. What is visible: the triangle chalked on the floor and the trace of the small lamp on the top of the head of the person being tested. The arrow indicates the direction of movement. The task was executed well. (b) The same experiment but with another person being tested, whose ill condition is well reflected in the photo. The light dots above are the person being tested and the researcher, taken from a 'bird's-eye view'. Both photos were taken by the author with Dr. N. Ozeretskii[6]

---

[6] Nikolai Ivanovich Ozeretskii (1893–1955) developed a method for investigating motor activity in 1923, which was named the 'Method for mass evaluation of motor activity in children and adolescents', which is still used today.

Lecture 10

Finally, Lecture 10 compares two variants of the hammer strike, analysing each in terms of biomechanics as an example that can be extended into other movements.

Comrades! As yet, there is no other labour movement available which has been developed in my laboratory with such precision and detail as the hammer strike. The reason is that the research that I was conducting on labour movements in TsIT ceased after my resignation and I went on to the study of movements in quite a different area. Therefore, what is appropriate for me to offer you today as a model for the examination of exemplary movements is in fact the strike; as for other types of labour actions, it would be more convenient for us to study them by way of questions and answers or by seminar discussion in the workshop.

Let us begin with the hammer strike with one hand. This is a movement, which appears quite uncomplicated, as if there is not much to say about it. In fact, the situation is not so. The movement of the swing together with the strike takes up about a second of time; during the working day, the professional carpenter is able to carry out no less than 10,000 strikes. If he expends in each strike in incompetent work even minimally superfluous effort, then from 10,000 such expenditures a quite impressive sum mounts up, which can and must be avoided.

In order to interest you at once in the question of the correct strike, I will ask you how you, strictly speaking, move your right hand in chiselling and how you learned to perform this movement? Describe or show it.

Student: Here, doctor, we all want to ask you how the strike should be done. Here among us there are lads who are experienced carpenters and they always move their elbows outwards; but now we've been shown several times to chisel by moving it straight in front, but then the hammer falls back. So we don't really understand which is more correct; it seems somehow awkward to take it forward.

Lecturer: And which do you think is more correct?

© Springer Nature Switzerland AG 2020
N. A. Bernstein, *Biomechanics for Instructors*,
https://doi.org/10.1007/978-3-030-36163-1_11

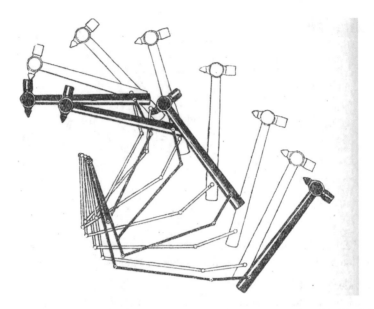

Fig. 10.1 The consecutive positions of the right hand with the hammer during chiselling. The long axes of the upper arm, forearm and hand are conventionally depicted with straight line links. The white links and the hammers are the swing-up, and the black—the strike. The difference in time between the adjacent positions is 1/15th of a second. The strike is near to normal 8

Second student:	The lads who have trained at TsIT were also told to chop in one plane but then you do not get the force. If you take it out from the body, it is somehow easier.
Third student:	Here, try it; learn to take it out from the body and you will injure all your fingers. You'll never get the accuracy. (*Lively debates*).
Lecturer:	So this question is interesting for you?
Students:	Yes it is.
Lecturer:	So here is the start of today's lecture. We will investigate which of the strikes is the more correct, in a systematic way.

I have recorded probably many hundreds of cyclograms of strikes. So we have a big choice. But what is interesting is that the variety of different types of strike turned out to be insignificant. In essence, I had to do only with two main forms; all the rest were varieties. Now I will present you with sketches which portray the consecutive positions of the right hand and the hammer in two typical cases (Figs. 10.1 and 10.2). Look at them and try to work out what type of strike each of them portrays.

Students:	(*Indicating* Fig. 10.1): This one is of the move to the side.
Lecturer:	I see that you have got your eye in a little. I have no other sketches like these, but there are enough cyclograms. So to begin with, let's relate in order everything that is clear on these two sketches, in order to find out how to recognise easily the details on the untreated cyclograms later. Let us begin with the strike to the side (as we will call the strike in Fig. 10.1) because biomechanically it is simpler.

Fig. 10.2 The consecutive positions of the arm with the hammer during incorrect (vertical) strike. The designations are the same as in Fig. 10.1

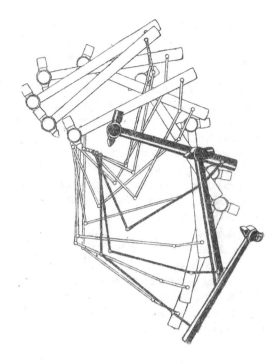

You must have the indubitable impression that the strike to the side is constructed more simply than the vertical strike. Quite simply, it is easier to investigate even in the drawing of the first than in the drawing of the second. In both drawings the swing-up is depicted with white strips and hammers, and the strike—with black ones. You see that in both drawings in the swing-up movement the hammer goes higher than the in the strike movement. This is something that is inevitable, which will have a place in each movement of the hammer and in any type of strike; consequently the trajectory of the head of the hammer will always have the same characteristic form, similar to a fish; in all further cyclograms you will recognise it easily.

There is nothing particularly noticeable about the path of the movement of the hand in either movement; you might only notice that in the strike to the side it is much shorter than in the vertical. In the strike to the side, it scarcely reaches the trajectory of the hammer and then in the vertical one, it intersects it and goes even higher above. Perhaps you will see this more clearly in Figs. 10.4 and 10.5. In connection with this, the hammer in the greatest swing-up (which I will call raising the hammer) is organised almost horizontally in the strike to the side and in the vertical is thrown back far behind. You can easily see that in the transfer from the swing to the strike the hammer must complete a much bigger turn in the vertical strike than in the one to the side.

The path of the elbow is very sharply distinguished in both types of strike. You will see this best of all in the actual cyclograms. It is so typical that with one look at it on the cyclogram you will always easily establish what kind of type the strike was. Here you see. The path of the elbow in the strike to the side is a little oval, inclined backward; in other words, the elbow moves back and up. Since the forearm which you see in Fig. 10.1 moves very little, the movement of the elbow up can in actuality be only the result of the movement outwards; what Fig. 10.1 shows is in fact the movement in the plane, which is in fact spatial movement, that is, it is diverted from its depth. The movement of the elbow in the vertical strike has also the characteristic form of a horn, with the point sticking up and forward. The swing of the elbow in the second case is much greater than in the strike to the side. Finally, the movement of the upper arm is also greater in the vertical strike than in the one to the side, as can also be seen in the same drawing.

Thus the displacement of all parts in the vertical strike is in general greater. The swing is less with only one subject—and indeed with the hammer. The arm sweeps more in the vertical strike and the resulting movement of the instrument turns out to be less. This is the summary of the first observations.

Let's express in numbers what has been said in words until now. You can take such a calculation as a general rule. The final velocity of the hammer head before the strike in the correct movement must be approximately as many metres a second as the swing of the hammer head has decimetres in length. Thus, the normal strike velocity of the hammer head in a swing of 80 cm will be 8 m/s. In our example, the length of the trajectory of the hammer head in the vertical strike (Fig. 10.2) is 58 cm, and the striking velocity is 6.4 m/s; in the strike to the side (Fig. 10.1) the length of the trajectory is 67 cm, and the strike velocity is 8 m/s. Consequently, both the length of the path of the hammer head and the velocity of the strike (and consequently also the force of the strike) in our example of the vertical strike are 20% less than in the strike to the side. Let us see what the cost of this was in both one and the other example.

The path of the hand of the arm with the strike to the side is 53 cm and in the vertical 75 cm. The path of the elbow in the first is 13 cm and in the second 23 cm. However, this means that the lesser effect of the strike of the vertical type demands a 30–40% greater swing of the arm. We can observe one more circumstance. With a strike of one hand usually 0.9 of the whole strike speed is a result of the muscles and only 0.1 is due to the hammer's fall from a height. In chiselling, the hammer is not raised very high; but once it has been raised it is necessary to be able to use all the work at full power which it can complete, when it is lowered back. Here are the figures showing how the raising takes place in this and the other type of strike.

In both one and the other it is necessary to raise to some height both the hammer and the centre of gravity of the arm. Since the raising of the hammer is fully used up when it falls back down, but the lifting of the centre of gravity of the hand is not used at all, evidently it is more advantageous to construct a movement so that the centre of gravity of the arm has been raised as little as possible and the hammer at the same time as much as possible. So here, in our example, the centre of gravity of the hand is raised to 25 cm in a vertical strike and the hammer at the same time to 21 cm. In the strike to the side the centre of gravity of the arm is raised to 18 cm and the hammer to 27 cm. In the first case, there is a loss of 4 cm, and in the second a gain of 9 cm. With all this we have already some data for orientation.

Let's go on to speeds. And here the principle of reasoning remains the same; the aim of the strike consists of the achievement of a certain speed of the hammer; all the rest are in essence the auxiliary adaptations, which it is necessary to construct as economically as possible. Let's put it like this: everything that is not directly necessary for imparting speed to the hammer is harmful. From this point of view both the swings and the speed of movement of the arm must be as small as possible, since they are not of direct use. The swing and the striking speed of the hammer must be as great as possible but the swinging speed must not be very big, since no direct use comes from it. Let's now review here the speeds of both types from this point of view.

Look at Fig. 10.3. It shows the movement speed of the strike to the side. The left side is the swing-up and the right the strike. Look first at the strike. The continuous curve is the speed of the hammer. The dashed curves are the speeds of the joints of the arm. You can see that even the biggest of them is more than twice less than the hammer speed of the strike. In the figures the speed of the hand is here equal to 3.7, but the speed of the hammer is 8 m/s. If you look at the swing-up you will see that here the speed of the hand everywhere is much lower than that of the hammer too. And also the swinging speed of the hammer is almost three times less than its strike speed. Economy is here clearly observed.

And here are the corresponding figures for the vertical strike. The strike speed of the hammer is 6.4 m/s, and the corresponding speed of the hand is 5.1 m/s. The hand moves a little more slowly than the hammer; it is not known why this is. The speed of the hammer is much less than in the strike to the side, at the expense of an extremely much higher speed of the hand.

Fig. 10.3 The speeds of movement of the parts of the right arm with the hammer during the correct strike (chiselling): _____ speed of the centre of gravity of the hammer; ------- speed of the fingers; .... speed of the wrist; ⸺·⸺· speed of the elbow. Further below—tenths of seconds; the divisions on the left—metres per second; from 0 to 0.4 s is the swing-up, from 0.4 to 0.6 the strike

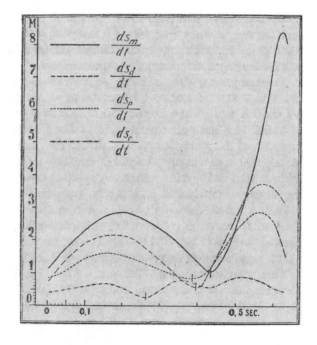

Fig. 10.4 Cyclogram of a
strike to the side, close to
Normal 6. Photographed
by the author at TsIT

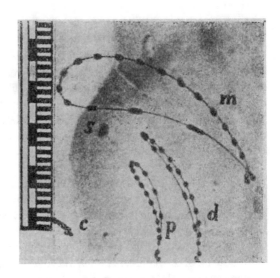

Let's take another strike to the side, a lesser one (Fig. 10.4), where its striking speed approaches that in our example of a vertical strike. We will get the following ratios.

	Strike to the side (m/s)	Vertical strike (m/s)
The striking speed of the hammer	6.3	6.4
The swing-up speed of the hammer	1.7	3.0
The greatest speed of the hand	2.6	5.1

The same ratio, as it turns out, appears in all comparisons of these two types. With the vertical strike there are always extreme speeds in all those places, where no direct use comes from them. The distances covered by the limbs also turn out to be more than necessary during the vertical strike.

Let us stop our systematic investigation of the two strikes for a while and occupy ourselves with the cyclograms. I want you to get the idea straightaway that I have not selected figures and events deliberately. Let us begin with the vertical strike.

In Fig. 10.5 one more typical vertical strike is shown. Now you know already how to read the cyclogram on first glance. You see here some features that are already well known to you. The path of the elbow is that of the same original horn; the hand is carried high in a similar way—its path consists here of more than 50 cm; the hammer similarly falls back deeply. The path of the hammer is small again (44 cm); the striking speed (5.1 m/s) is also small. You will clearly remember that the speed of movement is greater the more sparsely the points on the cyclogram are situated. So here in Fig. 10.5 it is evident that the striking speed of the hammer is almost no greater than the greatest speed of the hand.

Here is another example of the vertical strike (Fig. 10.6). Perhaps you will also recognise the face of the person who is photographed here. In this snapshot, again all the details of the vertical strike are known to you; give your attention only to the

Fig. 10.5 Cyclogram of
the vertical strike.
Photographed by the
author at TsIT

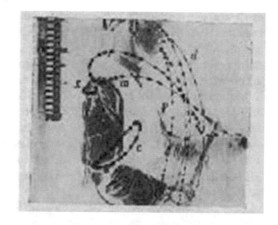

Fig. 10.6 Cyclogram of
chiselling. The vertical
strike with a big recoil.
Photographed by the
author at TsIT

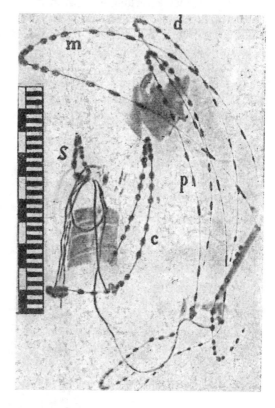

great tails which come down from each trajectory at the moment of the strike below, about 30 cm. Can you guess what this is? It is the recoil, the flying off of the hammer below after the strike. (*Exclamation of the listeners: it is true, yes, it also now strikes like that!*) Let us clarify whether this is a good thing or not? In the first place, we have here a superfluous movement, the extension of which back and forth reaches 60 cm; it is already uneconomical. In the second place, the jump back of the hammer takes place when the strike has reached the chisel not along its axis but obliquely. In these cases not all of the living force of the hammer is used for the strike; a significant part of it goes on the jumping back of the hammer; this part is the greater when the inclination of the direction of the strike to the axis of the chisel is the more sizeable. All this part evidently is lost in the strike; moreover, it is still necessary to brake it with the hand. Consequently, I strongly recommend that you avoid such recoils and oblique strikes, however elegant they seem to you.

I will add to this that this same subject gave me a completely different cyclogram a year before this, very much like the strike to the side. Since that year he, evidently, has retrained himself according to the TsIT standard, without it being particularly useful to him.

Here is a completely remarkable cyclogram for you (Fig. 10.7). You won't understand it immediately. I took it of one particularly zealous pupil, evidently, at that time when they were teaching the vertical strike particularly zealously at TsIT. The letters will help you investigate this cyclogram; *m* signifies the path of the hammer, *d* and *p* the path of the hand and *c* the path of the elbow. The last is the same as always.

In the first place, you see here a fantastically deep start of the stroke. The impression is such that it is as if the person is preparing to hit himself on the back. Here the path of the hand is 68 cm and the path of the hammer 39 cm in all, that is, almost half. The speed of the hand is also very significant; the greatest speed of the hand is

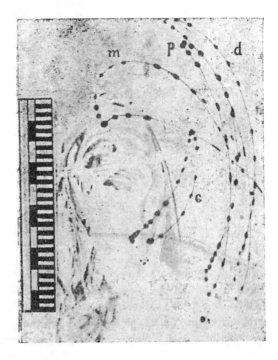

Fig. 10.7 A cyclogram of chiselling. The vertical strike with an excessively deep raising of the hammer. Photographed by the author at TsIT

the same as the striking speed of the hammer. But the most amusing thing in this cyclogram does not strike the eye immediately. Turn your attention to the place in the trajectory of the hammer where it goes from the swing-up into the strike, that is, the moment where it is raised. In the whole cyclogram, as in the majority of the rest, the points go in groups of three. But in this particular place between two sets of three there is only one fat point. What does it mean? No more or less than that the hammer has stopped here for the time in which three points have elapsed. From the outlining of the path of the hand in the corresponding time, it is clear that the hand has time to go down by a good 15 cm in the time where the hammer hardly moves at all.

Here there is one more superfluous movement that is concealed, which is much more ominous than what we have noticed with the recoil of the hammer from the chisel below. It is ominous because it has been observed already in all the vertical strikes without exclusion.

The task of the striking movement is, in fact, to convey the acceleration of the hammer head. It is clear that if any movement of the hand is completed with a fully or almost immobile hammer then this movement will not be of any use. Indeed this is what has happened with the strike that was photographed in our Fig. 10.7, but perhaps it could be suggested that what's at fault here is not the striking method but the lack of expertise of the person being tested. It turns out that this is not the case.

Look at all the pictures of the vertical strike which I have shown you and you will note in all of them one general phenomenon. The hammer begins to acquire speed, to gather acceleration in its striking path only when the axis of its handle is placed at a tangent to the direction of movement, that is, when the hammer begins to move along itself. In my essays about the strike, I gave a mechanical explanation of this phenomenon; I cannot repeat it here. What is important is that it manifests itself as a law with unchanging constancy.

As you see, on all our cyclograms the striking part of the path of the hammer begins almost horizontally, only rarely with a small slope from below to above. Therefore the hammer will be placed along the tangent only when its handle is almost horizontal. If the hammer was thrown far backwards at the beginning of the stroke, then the whole part of its movement up to the time when the handle has become horizontal falls away uselessly. During this time, it cannot mechanically gather speed, as it turns out. This is particularly clearly shown in Fig. 10.7; here it simply stands, all the time, until the handle becomes horizontal. In other photographs it moves in the same place, but always slowed down, and the points of the cyclogram in this place are denser.

The conclusion is this: the whole part of the swing, connected with the steep slope of the hammer backwards, and the whole part of the striking movement, returning it back into the rational position, represent an unnecessary loss of time and strength; they do not give a mechanical result, come what may.

From this point of view, it is interesting to compare the consecutive positions of the hand with the hammer with this and the other type of strike, as they became clear in Figs. 10.1 and 10.2. Both pictures confirm the rules that have now been stated. As you remember, in both, the striking movement is depicted with black, and the movement of the swing-up with white. The beginning of the striking movement is defined by the sign just pointed out. In both pictures, the interval of time between the two consecutive positions is 1/15th of a second. In the first place, it is quite evident that in the strike to the side the hammer is not carried up further than to the horizontal position. Consequently the striking movement can begin immediately after the swing-up from the furthest point of the trajectory. In opposition to this, Fig. 10.2 (the vertical strike) automatically gives the impression that it would be better if the hammer stood immobile at the raising up as it does in Fig. 10.7, than to move forward, uselessly shortening in this way the striking path that is possible for it, which is already small. In fact, in the diagram Fig. 10.2 it is clear how short the active part of its path is. Even with a very great force applied to the hammer during this, that is, with significant acceleration, it does not have time to make up a big speed in such a short path. This is what usually happens.

Let's turn now to the cyclogram taken in the cases of the strike to the side. I already showed you two of these (Figs. 8.13 and 10.4); because of lack of space I will limit myself to just one more, showing a strike that is not strong—about 4 m/s (Fig. 10.8). You can see their general characteristic features on all three cyclograms. I will recount them in brief. Everywhere, the size of the trajectory of the hammer (it is always indicated by the letter m) is much greater than all the rest of the trajectories. Universally, the trajectory of the joints of the arm (as with the hands d and p, so with the elbow c) is inclined back, not up and not forward as in the vertical strike. Universally, the speed of all these joints is much less than the striking speed of the hammer. In the hammer itself there is a sharp difference between the speeds of the swing-up and the strike. The whole movement produces the impression of something concentrated on one basic aim—the striking end of the movement of the hammer. This economy of speeds, shown in the cyclogram by small gaps between the two

Fig. 10.8 A cyclogram of chiselling. The strike approaches normal 4. Photographed by the author at TsIT

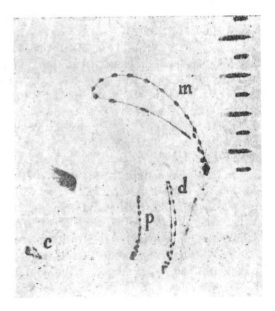

consecutive positions, is visible in the diagram (Fig. 10.1) in a particularly pictur-esque way. Here one single gap is great, corresponding to the movement of the hammer right before the strike. It alone has absorbed into itself a comparatively great speed and so it seems that all the remaining parts of the movement are econo-mised into its use as much as possible.

Is it possible then that all vertical strikes are so hopelessly bad? Regrettably this is so. Is it possible then that all the strikes to the side are irreproachably good? Well, no, it is impossible to say that. I shall reveal a little secret to you. I have shown you all kinds of vertical strikes and I have chosen only the best of those to the side that came out as a result of detailed investigations. Those three strikes to the side that you have been shown today are in essence three representatives of that type of strik-ing movement which I proposed as the biomechanical normal of the strike. There is very little use in criticising and condemning that. It is much more important to propose something definite, undoubtedly good, which can be aimed for with instruc-tion. Here is the biomechanical normal that we developed and published at the same time as a confirmed result of our work on the biomechanics of the strike.

A different force of the strike is demanded in different situations from the strik-ing movements. Since the known force of the strike is at the same time its aim, then it is natural to classify them by their force. And since the force, in turn, depends, as we have seen, on the striking speed, we have decided to label various strikes accord-ing to the value of their striking speeds. The interrelationship between striking speeds, the weight of the hammer head and the living force of the strike are defined in the following table.

Speed group	Striking speed of the hammer head	Work of one strike (in metre-kilograms of weight)			
		Hammer head of 200 g	Hammer head of 400 g.	Hammer head of 600 g	Hammer head of 780 g
2 {	2 m/s	0.051	–	–	–
	3	0.114	0.183	–	–
4 {	4	0.204	0.327	–	–
	5	–	0.509	0.765	–
6 {	6	–	0.734	1.101	–
	7	–		1.500	1.950
					2.545
8 {	8	–		1.958	3.223
	9	–			

As this table shows, it is possible to obtain very varied values for the living force of the strike, from 1/20th up to 3 m/kg and more, combining in a suitable way the weight of the hammers and striking speeds.

Now I shall explain what the first column of the table means: the speed groups. To describe the normal in conformity with all possible striking speeds would of course be impossible. At the same time it is clear that the 2 m/s movement is con-structed differently than the 8 m/s movement. That is why we have chosen as a description of normals four exemplary strikes that must give the final striking speeds corresponding to 2, 4, 6 and 8 m/s. All these normals have received the correspond-ing numbers: Normal No. 2, Normal No. 4 and so on. Is this understood?

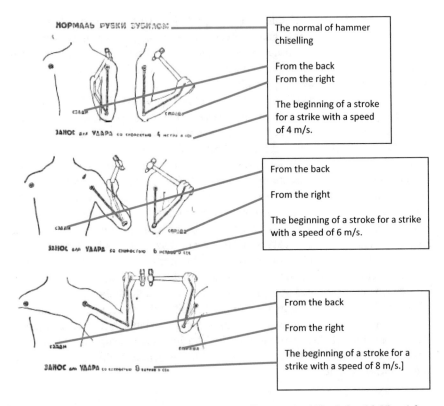

НОРМАЛЬ РУБКИ ЗУБИЛОМ — The normal of hammer chiselling

From the back
From the right

The beginning of a stroke for a strike with a speed of 4 m/s.

ЗАНОС для УДАРА со скоростью 4 метра в сек.

From the back

From the right

The beginning of a stroke for a strike with a speed of 6 m/s.

ЗАНОС для УДАРА со скоростью 6 метров в сек.

From the back

From the right

The beginning of a stroke for a strike with a speed of 8 m/s.]

ЗАНОС для УДАРА со скоростью 8 метров в сек.

Fig. 10.9 The raising of the arm and the hammer according to normal No. 4, 6 and 8. View 1 from behind and 2 from the right

Figure 8.13 is the movement that is most applicable of all to normal No. 8 out of all the strikes I have photographed. The diagram in Fig. 10.1 corresponds to the same normal. Figure 10.4 is close to normal No. 6 and finally Fig. 10.8 to normal No. 4. If you wish to examine them more attentively then you will find in them both much that you know and much that is evident from the picture better than from any detailed description.

The basic idea of all the normals is the combined throw of the upper arm and hand. In the smaller normals (No. 2 and 4) in the place of the upper arm is the forearm. In the bigger ones (No. 6 and 8), the distribution of roles in general is such that the main striking movers are concentrated in the shoulder and wrist joints, while the elbow and finger joints take on themselves for the most part the management of the purpose mechanisms, that is, precision of movement and accuracy (Fig. 10.9).

The swing-up is executed by means of shoulder extension.[1] This movement at first takes the elbow almost backwards and then up. It is completed with the help of the deltoid muscle. Simultaneously with this, the turning of the shoulder around the

---

[1] See previous notes about terminology and flexion and extension of the shoulder.

long axis outwards and the moderate flexion of the elbow take place. At the beginning of the swing-up the abducting muscles of the hand tense slightly; the muscles which flex the fingers are completely free in the course of the whole swing-up. In the course of the swing-up a slight supination of the hand takes place too.

Beginning approximately with the middle of the swing-up, the hand begins to brake elastically the hammer's flight backwards. The braking takes place by means of the slow recruiting of the extensor of the elbow and latissimus dorsi; at the very end of the swing the tension of the adducting muscles of the hand begins.

At the moment of lifting or a bit earlier the large pectoral muscle is energetically engaged and at the same time the tension of latissimus dorsi gradually increases. As a result of their joint action the shoulder firstly begins to move slightly forward (in the direction of adduction), then forward and down, and finally down and back along the line of pure flexion. It must be emphasised that these directions are in correspondence with the greatest force that these two muscles can manifest. At the same time, thanks to the joint action of the great pectoral muscles and latissimus dorsi, the rotation of the upper arm around the long axis inwards takes place. At the same time, due to the joint action of greater pectoral muscle and broad muscle of the back the rotation of the elbow around the longitudinal axis inwards takes place too. In the elbow joint the extensor of the elbow tenses and in very big swings so does the flexor of the elbow. Finally, immediately after the moment of lifting the flexors of the three last fingers (from the middle finger to the little finger) are included.

Among the enemies of the strike to the side and the prophets of the vertical wisdom the conviction reigns that the hammer does not fall backwards during the strike to the side thanks only to the convulsive clenching of all fingers into a fist at the height of the lifting. This is completely untrue. The clenching of the fingers at the moment of lifting manifests itself in one and the same way on the cyclogram, with an acute angle by which the trajectory of the hammer ends in these cases and a little distance between the swinging and striking path of the hammer. There is none of all this present in the normals, nor in the majority of the strikes to the side, and each of you, if he is able to observe, notices that with the strike to the side his fingers do not tense one bit more than with the vertical one.

Towards the end of the strike, that is, the moment of the approach of the most crucial part of the movement, the following happens. The greater pectoral muscle and latissimus dorsi have by this time already done their kind of military artillery preparation; they have imparted to the main mass of the hand and its centre of gravity its basic beginning speed. Now on the background of this beginning speed the hand gets involved in the affair.

The hand makes a sharp throw, formed by adduction and a small share of the supination. This takes place by means of the sharp excitation of the adductors. At the same time, the speed acquired by the upper arm brakes with the help of the flexor of the elbow. Finally, the fingers increase a bit more their flexing tension.

It is interesting to investigate purely mechanically what happens in this last short interval of time (it continues for 5–8 hundredth parts of a second) before the strike. The arm has towards this time a certain quantity of movement, that is, acceleration. Now what lies ahead is to use this acceleration in the best way possible. You understand that the living force of the upper arm, the forearm, etcetera do not impart any direct use to the strike. For the strike one thing is necessary: the living force of the hammer and that is all. And here is the mechanical meaning of everything that takes place in the last moments before the strike and leads to the fact that the whole living force accumulated before it by all segments of the arm is immediately conveyed to the hammer.

As is well known, work and energy cannot be nullified; they can only be transferred from one body to another. As a consequence of this, it turns out that when the segments of the arm begin to brake their flight, their living force is conveyed to the hammer with the correct movement; they brake, and the hammer, on the contrary, gathers acceleration. You will discover what I have just said on all the cyclograms of the correct strike and also on many of the cyclograms of the incorrect ones. That particular last period before the strike that is expressed in the hammer by the final and decisive flight is accompanied by a sharp deceleration of all the segments of the arm. By the way, slowing the hand in this way is advantageous because of the fact that it makes the hammer turn quickly in the hand; and such a turn makes far better use of the working possibilities of the hammer.

Let us use the occasion in order to check through a muscular inventory for both types of strike. We have seen that with the strikes according to the normal the most responsible moments of the movement are completed in the line of the strongest action of the strongest muscles of the hand. In contrast, muscles in the vertical strike are used much less fully. The movement of the elbow and the shoulder in the swing is completed by means of the smallest muscles of the shoulder (coracobrachialis) and only a very small part of the deltoid muscle can be involved in this joint action. The lowering of the shoulder in striking happens by means of contraction of the greater pectoral muscle in a much less advantageous direction than in the condition of the normal. As a result of less use of muscular work what will inevitably result is a much weaker strike.

You have already seen from the previous situation that the advantages of our normals consist not only in the fact that in them the muscle inventory is more weight carrying. The construction itself of the whole movement turns out to be mechanically more convenient and more advantageous; and I would place the main stress here in fact on the construction.

I will not put before you step by step in this course the details and figures characterising our four normals. Firstly, they are described in detail in my article entitled 'The Biomechanical Normal of the strike' in the collected *Issledovanie TsIT* in 1924. In the second place, I generally do not like to overload a lecture with prescriptions and figures, which each of the students can assimilate from the book much more easily and successfully. I will tell you only this in relation to the first reason. If you get hold of the article which has just been mentioned, then boldly go past the first two chapters, where the mathematics might frighten you. Begin straightaway with the chapter 'Biomechanics of the horizontal strike'; from here to the end, the exposition is carried out in quite a comprehensible way designed for practical workers, like you.

Now a very quick glance at what we have learned in this course. I would say that we should stress three points from all those contained in it.

In the first place, we have to some extent got familiar with the structure, biomechanical possibilities and characteristics of the human machine. We have spent quite a lot of time on this. This is understandable since sufficient acquaintance with basic characteristics of the machine allows us in the future to adapt our knowledge to all possible eventualities.

In the second place, we have looked briefly at those methods which are now being used in the biomechanical study of movement and have learned from an example how to make use of one of them swiftly. I am afraid that we do not however have the experience and the preparation for studying movement independently; but at least you are gaining some general literacy in this direction which will allow you to critically relate to proposed teaching methods and analyse them to some extent.

In the third place, finally, we have traced one labour movement quite painstakingly and ascertained that it represents a particularly complicated and multisided collection of mutual interactions between muscles, joints, centres of gravity and so on. We have also taken some account of how the separate details of such a movement exist in such close dependence with each other and how a change in one place must inevitably bring about changes in all the rest. All this was done, of necessity, very briefly, because you cannot expect to exhaust the great area of biomechanics in ten lectures; but perhaps the few indications which I have managed to give you will be of some use to you and will make you if not engineers of the human machine at least its capable machinists.

Printed in the United States
by Baker & Taylor Publisher Services